GERMAN LIGHT FIELD ARTILLERY

July 28, 1941. Horse-drawn light artillery passes a burning Soviet tank as it moves to a new position.

1935 - 1945

JOACHIM ENGELMANN

Schiffer Military/Aviation History
Atglen, PA

The preparatory bombardment at Kursk on July 4, 1943 was undertaken by the infantry divisions, most of which were equipped with the leFH 16, like the 3rd battalions of light infantry divisions and mountain artillery regiments. The ammunition-gunners of this gun crew include a Russian auxiliary volunteer, who is catching the empty shell casings. Behind the piece a quantity of shells lie ready. In addition to cleaning rods, rammer, unloader and manual fuse setters, the leFH 16's accessories included drag ropes for moving the gun by hand and a lever. (BA)

Translated from the German by David Johnston

Printed in the United States of America.
ISBN: 0-88740-760-9

This title was originally published under the title,
Waffen-Arsenal Waffen und Fahrzeuge der Heere und Luftstreitkräfte Deutsche Leichte Feldhaubitzen 1935-1945 Arbeitspferde der Divisionsartillerie,
by Podzun-Pallas-Verlag, Friedberg.

Published by Schiffer Publishing Ltd.
77 Lower Valley Road
Atglen, PA 19310
Please write for a free catalog.
This book may be purchased from the publisher.
Please include $2.95 postage.
Try your bookstore first.

We are interested in hearing from authors with book ideas on related topics.

LIGHT FIELD HOWITZERS

From the 15th Century to the Second World War

Wars of the 15th Century. The Czechs called their 150-250mm field guns, which could deliver flat trajectory as well as high-angle fire, "houfnice." At that point the guns were still non-mobile. Technically-improved and mounted on wheeled carriages, by the beginning of the 18th Century howitzers were an important weapon of fortress and siege artillery, with a maximum firing range of 1,500 meters and a rate of fire of two shots per minute. The Prussians introduced a few field howitzers into their cannon batteries during the Seven Years War, while the siege artillery already had 25- and 30-pounder (220mm and 240mm) howitzers in addition to its 7- and 10-pounder (150mm and 170mm) weapons. In the first half of the 19th Century lessons learned in the wars of independence resulted in limited technical improvements in firing range, decreased weight and a more favorable relationship between mobility and effect. The Prussian artillery received a 25-pounder (226.5mm) with a range of 3.15 kilometers and a 50-pounder (284.3mm) howitzer with a maximum range of 3.2 kilometers. A 7-pounder howitzer then under test entered service in 1850.

With the development of the rifled barrel in Prussia in 1861, the significance of the field howitzer decreased for a period. Not until the end of the 19th Century, on account of the Russo-Turkish War of 1878 and the Russo-Japanese War of 1905, and realistically with a view toward the next conflict, was attention focused on the howitzer again. Technical improvements fundamentally changed the nature of the weapon: in particular recoil in 1889 and a recoil brake in the lower carriage in 1905, the new panoramic sight for indirect aiming in 1905, the shrapnel shell and impact and timed fuses. In 1899 every corps of the German Army received a heavy and a light howitzer battalion, and from 1912 on every division had a light howitzer battalion.

Exploitation of terrain, concealed positions and entrenched infantry produced a need for a gun with which to effectively engage such targets in the field in a direct way. Consequently, at the turn of the century every European army supplemented its artillery with light and heavy field howitzers. Germany and Austro-Hungary decided on a caliber of 100 millimeters, France, England and Russia on 120. Apart from light infantry guns, there never was a howitzer under 100mm. Smaller states purchased the developments of one of the two groups. In the end the caliber of 105 millimeters, since September 6, 1900 called the "light field howitzer," became the most widely used. A short barrel, significantly greater quadrant elevation and a steeper flight path with greater angle of impact and increased ground effect were characteristics of the howitzer. Its total weight was significantly less than cannon of the same caliber.

Firing range increased steadily, reaching about 6 kilometers before the First World War, 10 by the end of the war and roughly 13 in the Second World War. Today it is up to 15 kilometers. Range of traverse was equivalent to that of field cannon and was limited to a few degrees. True, designs were produced in Germany during the Second World War incorporating the all-round carriage. Invented in the United States after 1918, it almost doubled the weight of the gun. However these designs progressed no further than models, for they required motor transport. Between the two world wars the initial velocity of the projectile was increased by 20 to 30 percent, but at the cost of an increase in gun weight, which had to have an effect on cross-country mobility and means of transport. In order to achieve a variable flight path, howitzers required two-part ammunition, with varying propellant loads based on the position, nature and range of the target, as well as the desired effect on the target. The greater the propellant charge, the flatter the trajectory with increased impact force with limited depth of penetration into the ground. Additional factors were the various types of projectile and different fuses, the position of the observer relative to the flight path of his battery, weather conditions and the angle of site between the emplacement and the target area. Prior to the introduction of electronic computers, observed fire was a technical skill requiring much training and experience.

With a total field artillery strength of 11,300 guns, during the First World War the ratio of cannon to howitzers went from 3 : 1 in peacetime to 5 : 1 at mobilization and 1.5 : 1 at the end of the war. The light field howitzer had obviously proved and established itself as a standard gun. Not without good reason did the Versailles Treaty forbid artillery with a caliber greater than 105mm. Germany was limited to 204 77mm field cannon and 84 leFH 16 light field howitzers with 800 rounds of ammunition per gun, one battery per battalion, while everywhere else the light field howitzer was increasingly replacing the light field cannon. Between the wars the field howitzer was obviously considered a suitable and effective weapon.

The Second World War

In 1939, after years of production, there were ten times as many field howitzers (1,968) as field cannon in the German Army. The field artillery made up two thirds of the entire complement of guns. The 105mm and 150mm were the Wehrmacht's standard equipment.

On the other hand Great Britain and the Soviet Union settled on the 88mm and 122mm cannon-howitzers. Except in a few cases, the light field howitzer generally proved ineffective in an anti-tank role when faced by the masses of Soviet tanks, especially since this was considered a secondary role. This shortcoming was initially made good by the widespread use of the outstanding 88mm flak, an army development from the 1920s, then compensated for by upgunning the German anti-tank guns, assault guns and heavy tanks, which more and more became self-propelled guns. With 1,023 battalions, the light field artillery was three times as numerous as the heavy artillery, and all horse-drawn. In addition, there were 62 motorized light battalions belonging to the panzer-grenadier and panzer divisions, as well as the GHQ artillery and that of Germany's allies.

The Light Field Howitzer 16

The leFH 16 was developed from the Field Howitzer 98/09 by Krupp; introduced in 1909, it was a conversion of the Field Howitzer 98, which still had carriage recoil. Barrel recoil, which was invented by Heinrich Ehrhardt, brought about a revolution in artillery. The gun's box-trail carriage had a ground spade and a gun shield; at 1.58 meters the barrel was still short. Between 1,260 and 1,144 guns of this type were used in the First World War. Bulgaria and Turkey bought the same model. An almost identical intermediate type, the leFH 12/16 by Krupp, was purchased by Bulgaria, Romania and the Netherlands.

A second interim type by Krupp, the leFH 17, used the carriage of the 120mm Howitzer 12 sold to Switzerland by Krupp and brought with it four significant advances: an increase in barrel length to 20 calibers (2.098 meters), an increase in the weight of the gun to 275 kilograms, increase in muzzle velocity of 130-150 meters per second and a resulting 4-kilometer increase in maximum range to 10 kilometers. It was a trend in development that continued and drew closer to the characteristics of a cannon. In spite of this, however, only 294 of Krupp's leFh 17 were delivered to the field forces.

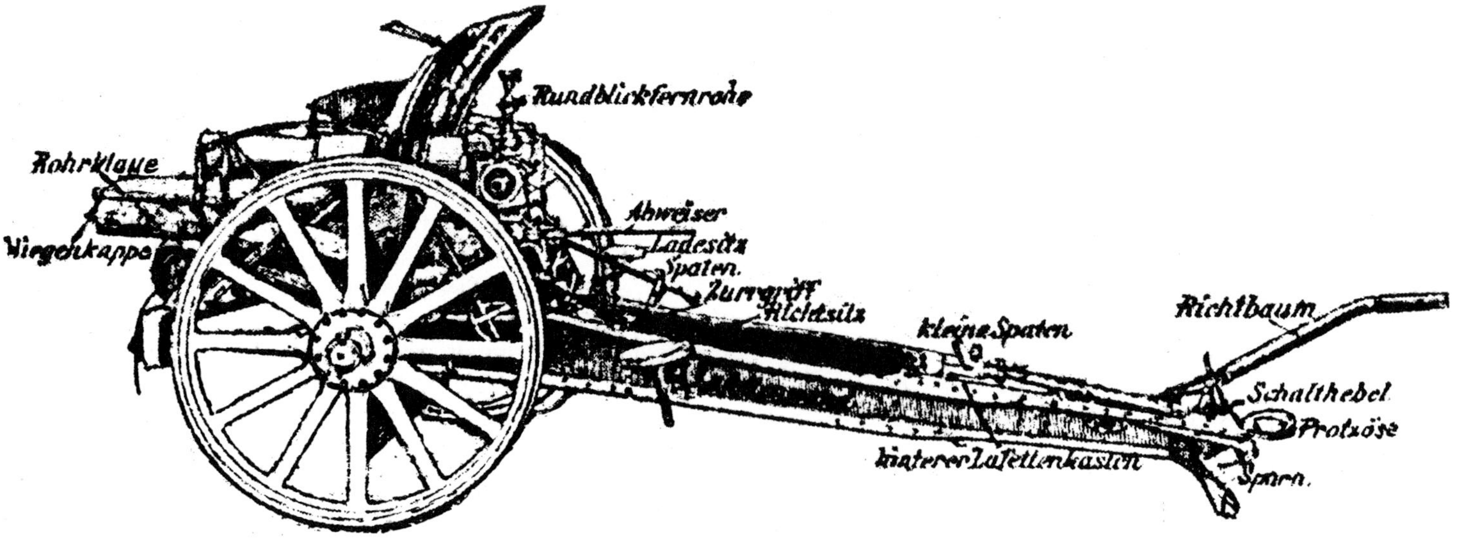

Field Howitzer 98/09, L/16, by Krupp. First 105mm field howitzer without carriage recoil, the leFH 16's predecessor. Introduced at the end of 1909, by 1914 there were 1,260 of these guns in the artillery battalions. In 1918 there were still 1,144 in service.

On the other hand the leFh 16 howitzer by Rheinmetall was widely used. It replaced the obsolescent leFH 98/09 completely, and by 1918 3,004 of these guns were in service on every front. Its barrel was also lengthened, to 2.29 meters, but weight in firing position was 120 kilograms less. As its muzzle velocity was somewhat lower, it could barely reach 9.7 kilometers, though with the C-projectile of the FH Granate 98. The slightly-tailed box-trail carriage and the retention of axle seats for the 1st and 2nd gunners, who rode with their backs to the direction of travel, were characteristics of this piece. Rheinmetall largely met the demands of the Artillery Examining Board held early in 1914. The first test battery was formed in the summer of 1916, and in September of that year all light field howitzer production was switched to the Rheinmetall weapon. In view of the military situation, as many components of the leFH 98/09's gun carriage as possible, as well as projectiles, cartridges and propellant charges, had to be used for the new gun carriage. Thus the leFH 16 was not a modern, new design, rather it was an interim solution which achieved high performance quickly, met all the requirements and, after some initial minor problems were eliminated, proved an outstanding success in the First World War. In the view of the experts the leFH 16 was unsurpassed until the end of the 1920s due to its excellent ballistic performance, great effect and adequate mobility. This excellent performance was partly due to increased projectile spin. It remained the standard howitzer of the divisional artillery until the introduction of the leFH 18 and was widely used in all theaters until 1945.

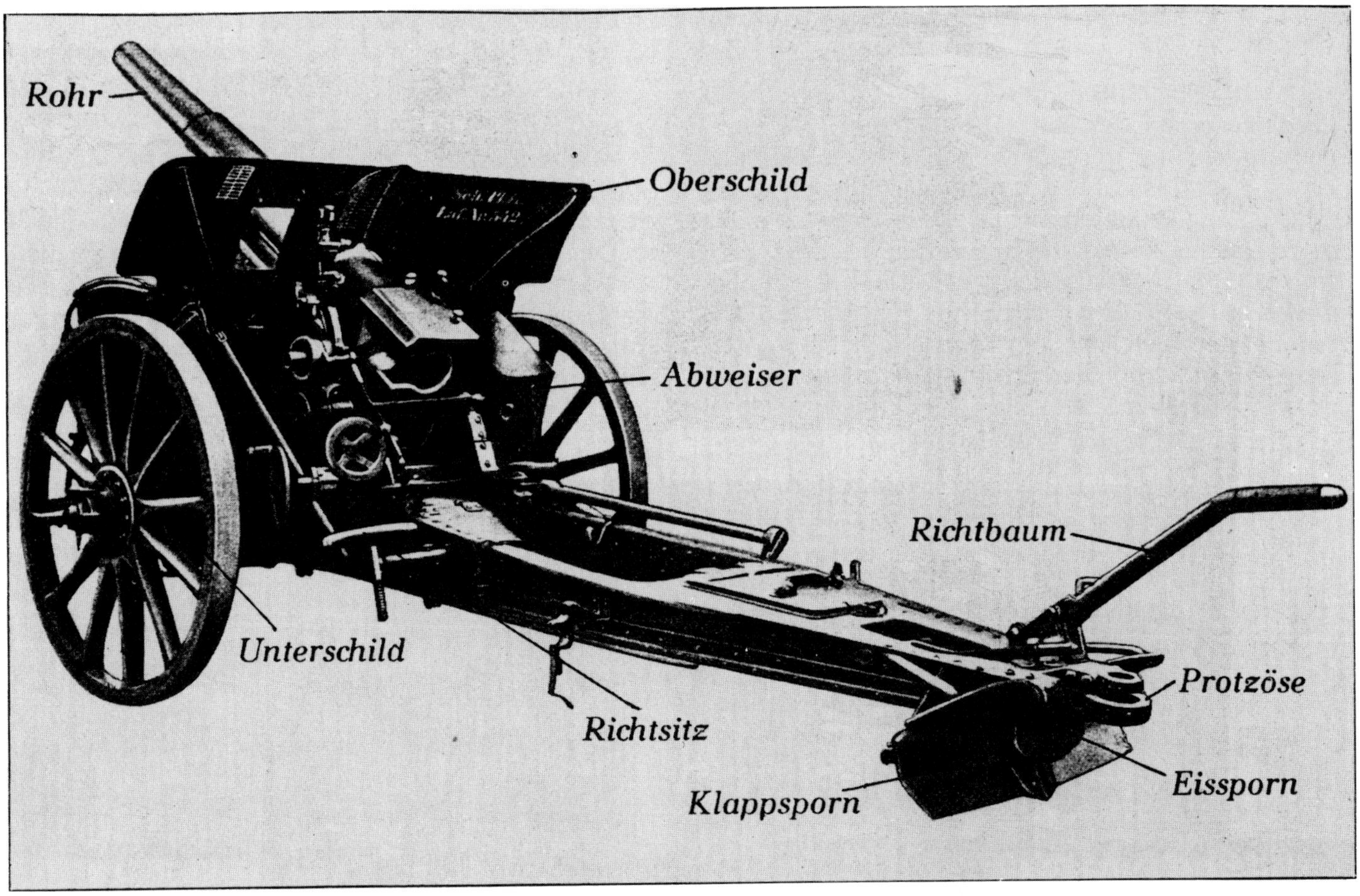

LeFH 16 by Rheinmetall. This view from behind illustrates all the gun's working parts. Barrel length was increased from L/16 to L/22, more than 60 centimeters.

A series of minor improvements resulted in the leFH 16 nA (new artillery). There were no design innovations apart from the breech mechanism, where the sliding wedge-type breech-block was replaced by a pusher-handle sliding-wedge type, because it allowed the gun to be served faster and its smooth surfaces were less susceptible to dirt. Instead the new model incorporated technical improvements in the

The battery's number three gun – full view from behind: the layer's seat with elevating and traversing mechanisms for the gun layer and the loader's seat for the 2nd Gunner, who fired the weapon. The cradle lock handle was used to lock the cradle with the barrel during transport to avoid damaging the compensator and the elevating mechanism.

lower carriage. Stiffeners were added to strengthen the gun carriage side plates. Counterrecoil and compensator springs were made larger; assembly guide claws, trunnions and trunnion bearings, elevating rack, outer elevating shaft and cradle cap were made more massive. In order to prevent the elevating rack from jumping out of the outer elevating shaft at high angles of elevation, a longer shoe was added to the rear of the cradle. An ice spade was added to the carriage tail in addition to the ground spade. A spring coupler was installed to keep the limber pole level and prevent an unequal burdening of the vehicle and thus the wheel team. There was also a rationalization of ammunition; the available types were the long FH shell with two kilograms of explosives, the Howitzer Shrapnel with powder-train fuse good only to 6,000 meters, the Howitzer Shell 15 with a 1.4-kilogram load and the C-projectile. Introduced in 1917, the latter's range could be increased to 9,700 meters by using the ninth charge, although it put a greater strain on the gun.

The leFH 16's box-trail gun carriage in detail – the trail handspike for lifting and turning, folding trail spade for soft ground and ice spade for hard ground, as well as the double lunette for the flexible limber connection.

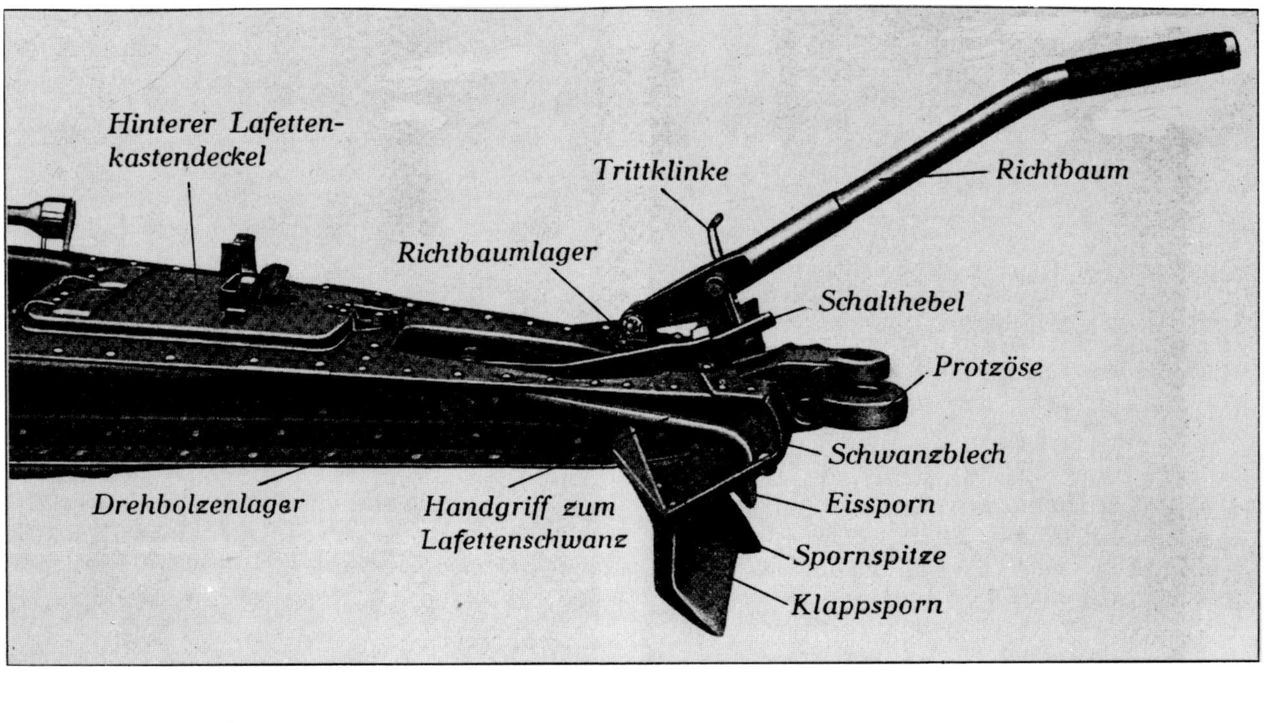

After the First World War the American magazine "Mechanical Engineering" gave the following rather general assessment of the leFH 16: "During the war the Germans easily took first place in the development of artillery. They developed and built significantly improved guns of all calibers, not only with regard to a more efficient barrel of a given caliber, but also to the installation of this barrel in carriages which enabled them to achieve the maximum firing range. By using these guns in the 1918 spring offensive the Germans caused the Allies great difficulties."

In 1921 the "Artilleristischen Monatshefte" (Artillery Monthly) expressed the following view: "Undeniably one is entitled to write a favorable report on the types of guns developed in such a difficult time and under the most difficult conditions. The guns met all expectations in regard to their ballistic performance and that remained the most important thing. All in all, taking all the conditions into consideration, after converting to the Type 16, which was born during the war, the bulk of the field artillery had been equipped with a weapon that made it possible to place the greatest demands on them, and the troops proved that with this gun they were in a position to live up to these demands." That the masses of allied artillery and their unimaginable quantities of ammunition, employed in a limited area, led to mastery of the battlefield in the First World War is another thing altogether.

At the time of the march into the demilitarized Rhineland on March 7, 1936, all German artillery battalions were still equipped with the leFH 16, which nevertheless represented an increase in caliber over anything possessed by the Reichswehr. Here the 2nd Platoon of the 5th Artillery Regiment's 6th Battery is seen in Rastatt.

In 1933 the Reichswehr had only 28 leFH 16 light howitzers in its 24 artillery battalions. By 1934 the number had risen to 496. There were 568 in 1935 and 728 in 1936. In 1937, the year the leFH 18 was introduced, the total number of leFH 16 howitzers was 980. Thus it was on the leFH 16 that the Wehrmacht's artillery cut its teeth.

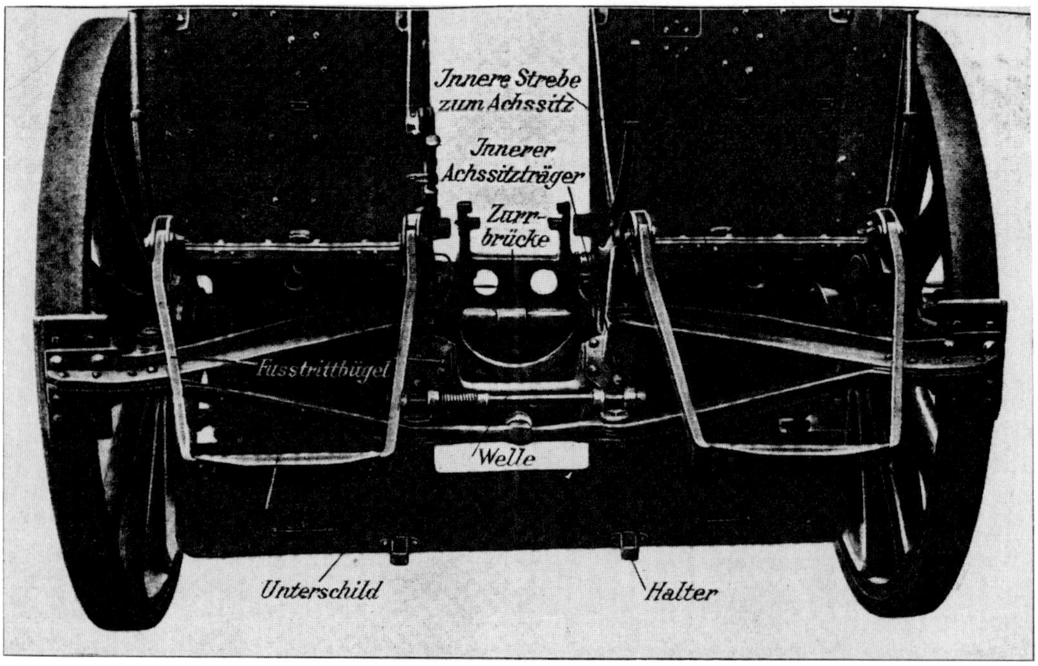

Left:
The axle seats on either side of the barrel, facing opposite to the direction of travel, belonged to the gun layer and loader. The gun carriage was braked from there when the six-horse team threatened to move.

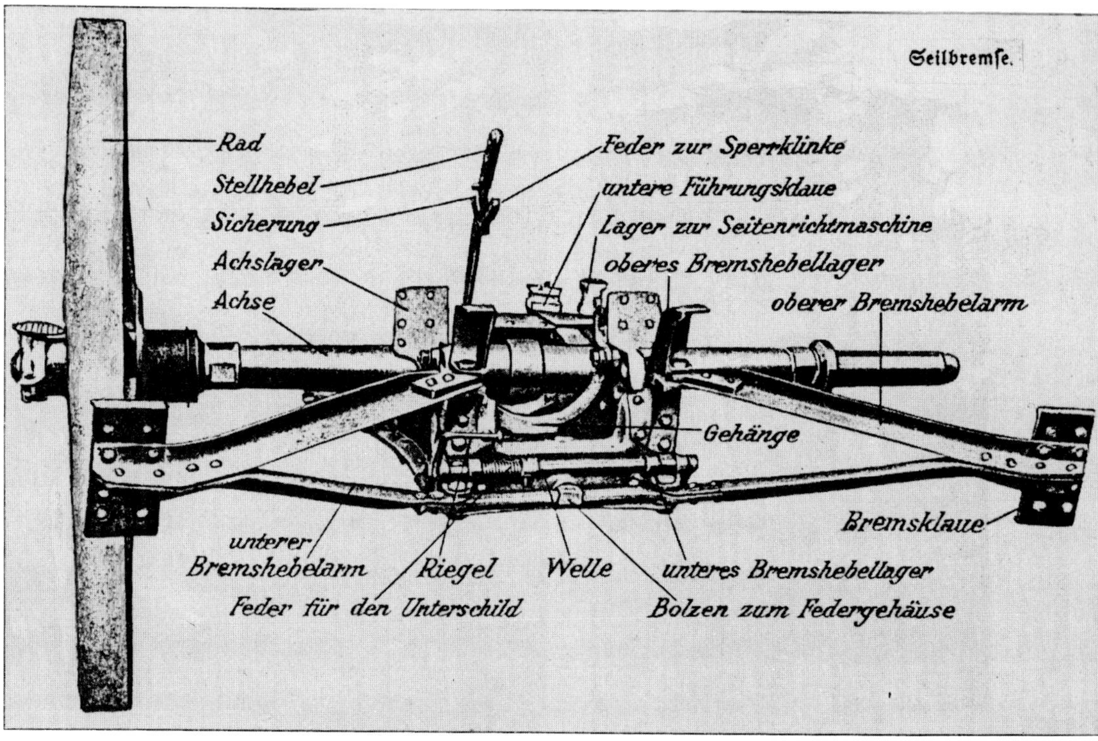

Right: While on the march, the cable brake of galvanized steel wire was operated independently by the gun layer from the axle seat by using the brake lever. This acted via cable drums and wooden brake shoes on the steel wheels, serving as a driving and firing brake. The wheels, which were one meter in diameter, had twelve spokes made of ash or elm, a mounting rim of mild-steel and on the outside the hot-mounted tire, held in place by twelve brads. The hubs were also made of steel.

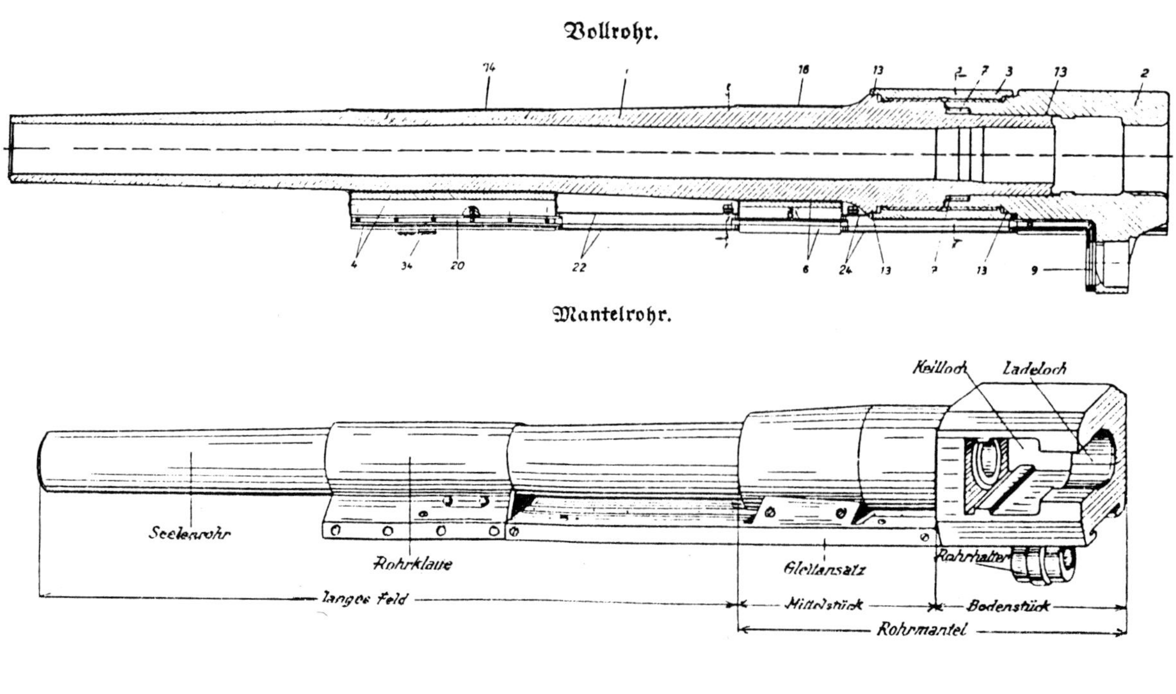

Left:
Like the built-up barrel, which consisted of a liner, barrel guide and jacket, the monobloc gun barrel with breech and breech ring was made of tempered chrome-nickel steel and burnished on the outside. The barrel guide held the barrel on the cradle's slide during firing.

The upper carriage traversed around a pivot in the lower carriage. The barrel and cradle rose and lowered with the trunnions in their pans. The sighting mechanism was fixed to the sight mount, which was graduated in elevation to 780 and which also had a meter scale graduation for engaging tanks using the Number 5 charge. It was moved by a high-speed drive, with a worm drive for fine adjustments in elevation. Supplementing this was the clinometer with level and the cross level with traversing mechanism for compensating for a poor alignment of the wheels, both with graduations.

A battery of leFH 16 light howitzers moves cross-country toward its firing position during the buildup leading up to the offensive near Kursk ("Citadel") at the beginning of June 1943. The trailers are fully packed, and the gunners are walking to ease the load for the horses. (BA)

The upper shield of 3mm armor plate, with its cutout for the barrel, sight port and sight port cover, was notable on account of its shield supports. The barrel cradle with the trunnions served as support for the barrel, the slide during firing, and the barrel brake housing which contained the braking cylinder and the three double spring recuperators for the barrel, with its 1.2-meter recoil. The recoil guard prevented the gun layer from being struck by the breech.

During quiet periods of coastal defense duty in Northern France in 1942, the battalion armorer-artificer official and his assistants from the batteries checked the guns. Here they are seen adjusting the panoramic telescope, part of the sighting mechanism and vital for accurate shooting. Coupled to the sight mount was the elevating mechanism, which had a range of travel of from -9 to +40 deg. On the leFH 16 the upper sight port cover had to be raised before aiming the gun.

Lappland, summer 1943. The gun section of an leFH 16 battery moves to a new position over rough, makeshift roads. In the foreground is the Number One gun. Seated on the left rear seat is the gun layer, who was responsible for the carriage brake. On the right is the loader. The crew had to protect themselves with mosquito veils in the damp northern woodlands. (BA)

Near Voronezh, June 1942. The left end gun of a 105mm M 14 (Skoda) field howitzer battery has just been unlimbered; the gunners unload the limber and prepare their gun for firing. The limbers were supposed to leave the position quickly. This howitzer was issued mainly to the Hungarians, Italians and Romanians. (BA)

The field howitzer's relatively light weight allowed it to be mounted rigidly on a stable sleigh; in this way it could be moved forward through dense forest and deep snow in order to take out enemy positions with direct or indirect fire. The sleigh was also more maneuverable in difficult terrain. Northern Lappland, winter 1942-43.

Lashed firmly to a circular base plate in a firing position dug in the meter-thick ice, a field howitzer stands guard over one of the many Norwegian fjords in 1943. The leFH 16 was indispensable in defending the endless coastlines and secondary fronts, especially since there weren't enough captured guns to do the job. (BA)

An leFH 16 in a field position in Southern France in 1944, prior to the invasion. The alert has been sounded, therefore the overhead netting and other camouflage has been removed, the muzzle cap has been taken off and the crew is ready to fire. While near the water line most – but not all – positions were made of concrete, farther inland the construction of positions was quite rudimentary. The batteries stood little chance against an enemy superior in all areas, especially in the air. (BA)

Cross-section and profile views of the new barrel for the leFH 18.

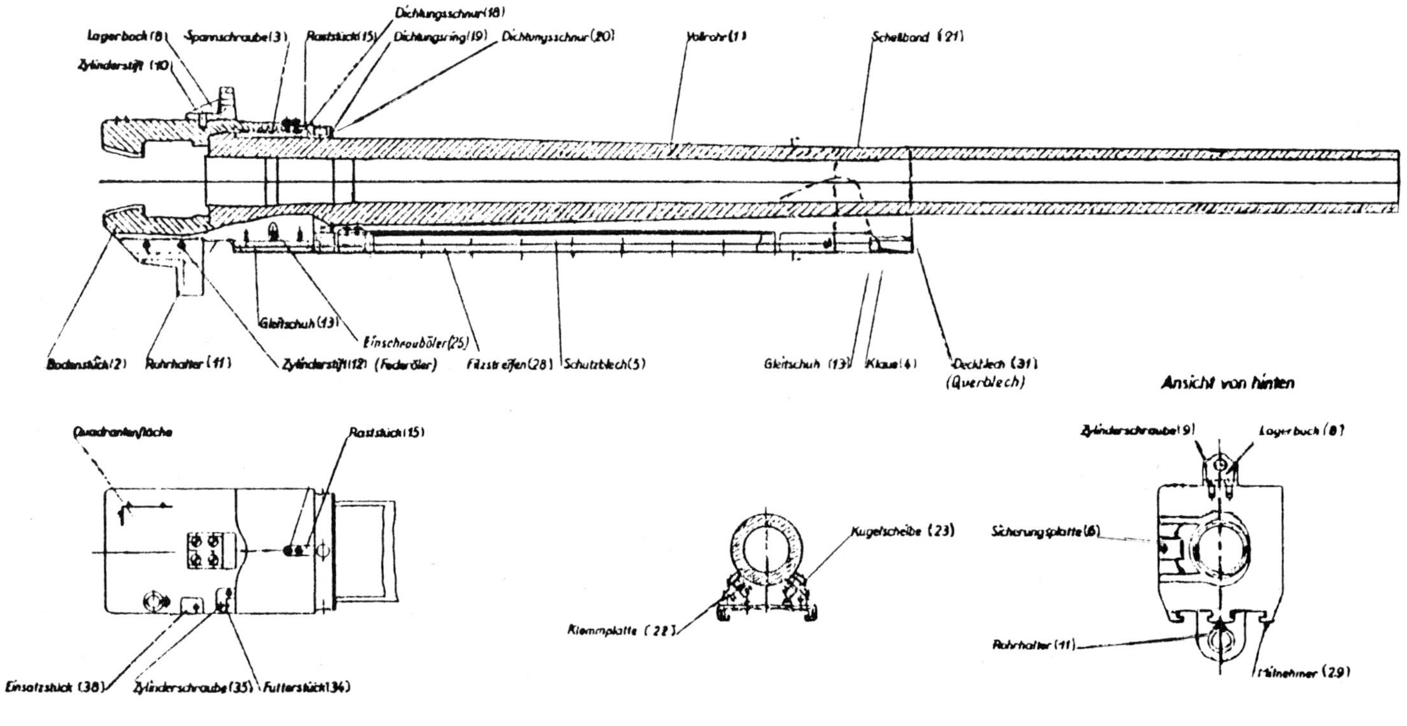

The Light Field Howitzer 18

Further development of the leFH 16 was requested by the Army Ordnance Office on June 1, 1927. In 1930 development work carried out by Rheinmetall in 1928 and 1929 led to the design of the leFH 18, first with a barrel length of 25 calibers, then 28. The new gun was in advance of its predecessor in almost every respect. Barrel length was increased by six calibers, or 63.1 centimeters, to 294.1 centimeters. Muzzle velocity climbed by 79 m/sec to 470 m/sec, with an increase of 1,400 meters in maximum range to 10.675 kilometers with the largest charge. The long-desired range of 10 kilometers had finally been exceeded! On the other hand the projectile was almost 1 kilogram lighter, with a slight increase in the size of the explosive charge to 1.845 kilograms. Completely new was the split-trail gun carriage with locking and large folding spades as well as carriage suspension. The axle was also equipped with shock absorbers, and in the event of a break it could be lashed up for slow travel (max. speed 8 kph) in an emergency. The three-point arrangement of the split-trail carriage made the gun significantly more stable, especially at higher muzzle velocities, and offered a tremendously increased range of traverse, 28 degrees to either side, particularly important in direct aiming. Light-metal cast wheels, 130 centimeters in diameter with 10-centimeter-wide hubs and steel ball-bearings, replaced the wooden-spoke wheels used until 1936. The metal wheels were more durable and made the gun easier to tow. Instead of the earlier cable brake they had a more effective internal expanding brake. Tractor-drawn versions were equipped with an additional air brake on account of increased speed and momentum. The metal casting process also simplified production compared to the earlier riveting, even though it increased weight. The new howitzer's protective shield was no longer rectangular in shape; instead it was built as a reinforced, tapered shield with flattened sides, extended by a folding lower shield beneath the 88-cm-high trunnion.

A rear view of the new leFH 18; in this case it is the tenth gun of a trials series built by Rheinmetall and the number two (B) gun of the trials battery. Note the longer slide, made necessary by the increased barrel recoil resulting from the higher muzzle velocity, and the automatic locking of the axle when the trails were opened. Even if the split-trail carriage did result in a considerable weight increase, the carriage's three-point arrangement was an invaluable advantage.

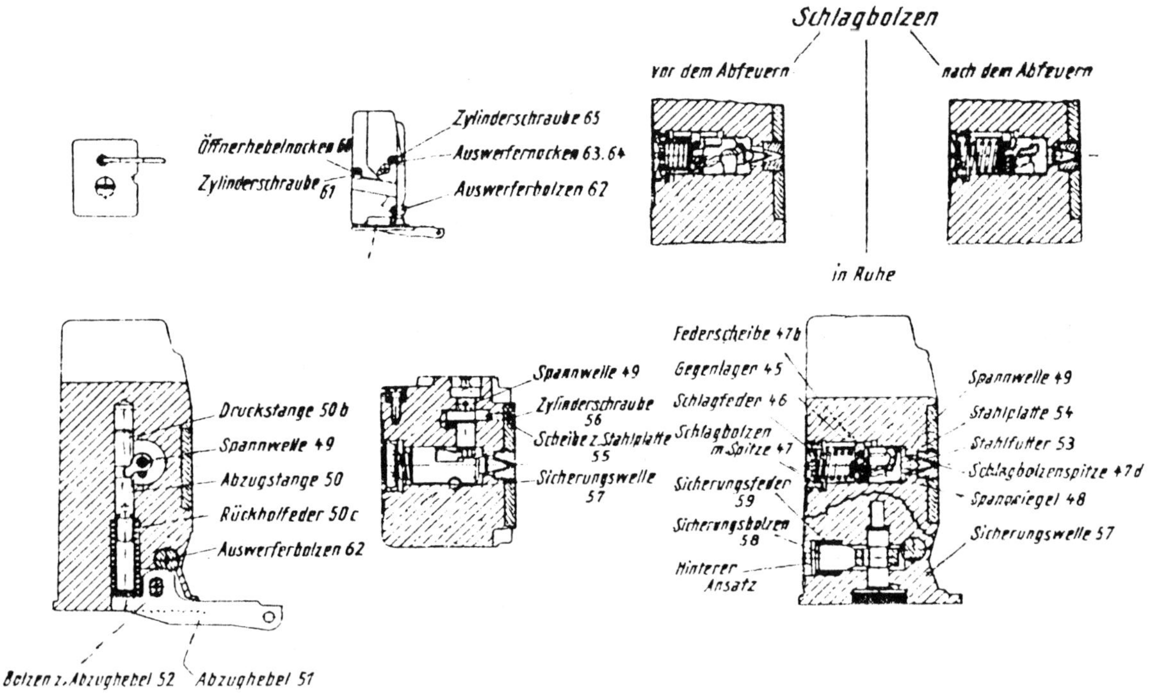

Outline of the mode of operation of the breech mechanism and firing pin.

The barrel brake (cross-section below, mode of operation on facing page) checked the recoil and provided the barrel's counterrecoil. It consisted of the braking cylinder, piston and piston rod, governor with counterrecoil piston rod, fluid equalizer and water jacket.

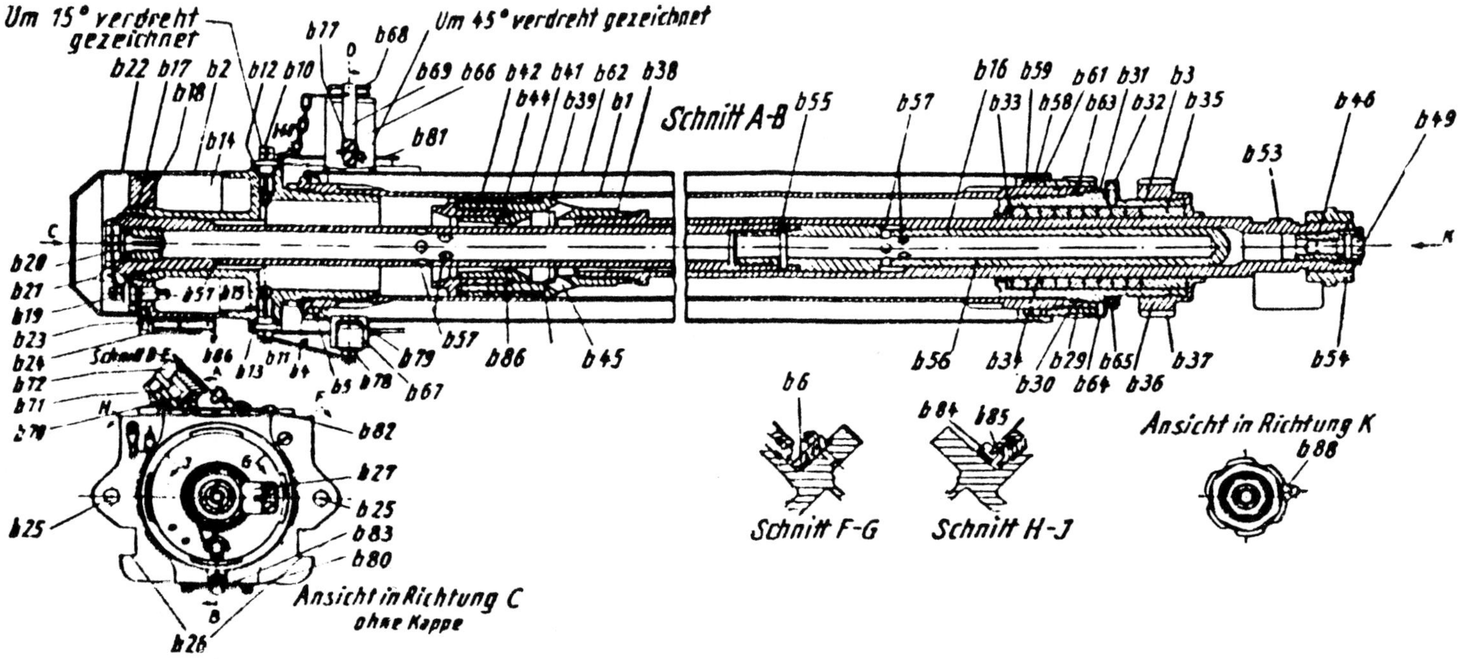

b

1.	*Braking cylinder*	*26.*	*Troughs*	*16*	*Control rod*	*69*	*Spring*
2.	*Fluid equalizer*	*27.*	*Safety lock*	*19*	*Filling plug*	*70*	*Valve stem*
3.	*Base bushing*	*29. 30.*	*Sealing ring*	*53*	*Cotter pin*	*71*	*Spring*
4. 5.	*Sealing ring*	*31.*	*Safety lock*	*54*	*Nut*	*72*	*Screw plug*
6.	*Safety lock*	*32*	*Cross-hole screw*	*55*	*Pintle*	*77*	*Screw*
10.	*Filling plug*	*33*	*Support ring*	*56*	*Longitudinal traverse*	*78*	*Screw plug*
11	*Bleed screw*	*34*	*Seal*	*57*	*Flow aperture*	*79*	*Gasket*
15.	*Pan*	*35*	*Bushing*	*58 60*	*Packing*	*80*	*Cross-hole screw*
16.	*Regulator*	*36*	*Pressure bushing*	*61*	*Thrust collar*	*81*	*Cover plate*
17. 18.	*Sealing ring*	*37*	*Teeth*	*62*	*Cooling water jacket*	*82*	*Screw*
20.	*Filling plug*	*38*	*Piston rod*	*63*	*Screw-down nut*	*83*	*Cover plate*
21.	*Sealing ring*	*39*	*Piston head*	*64*	*Safety lock*	*84*	*Cams*
22.	*Cap*	*11*	*Piston ring*	*65*	*Cross-hole screw*	*85*	*Lug*
23.	*Bolts*	*12*	*Guide bushing*	*66*	*Filler neck*	*86*	*Lever*
24.	*Compression spring*	*14*	*Brake bushing*	*67*	*Runoff socket*	*88*	*Safety lock*
25.	*Bores*	*15*	*Cutaway*	*68*	*Valve housing*		

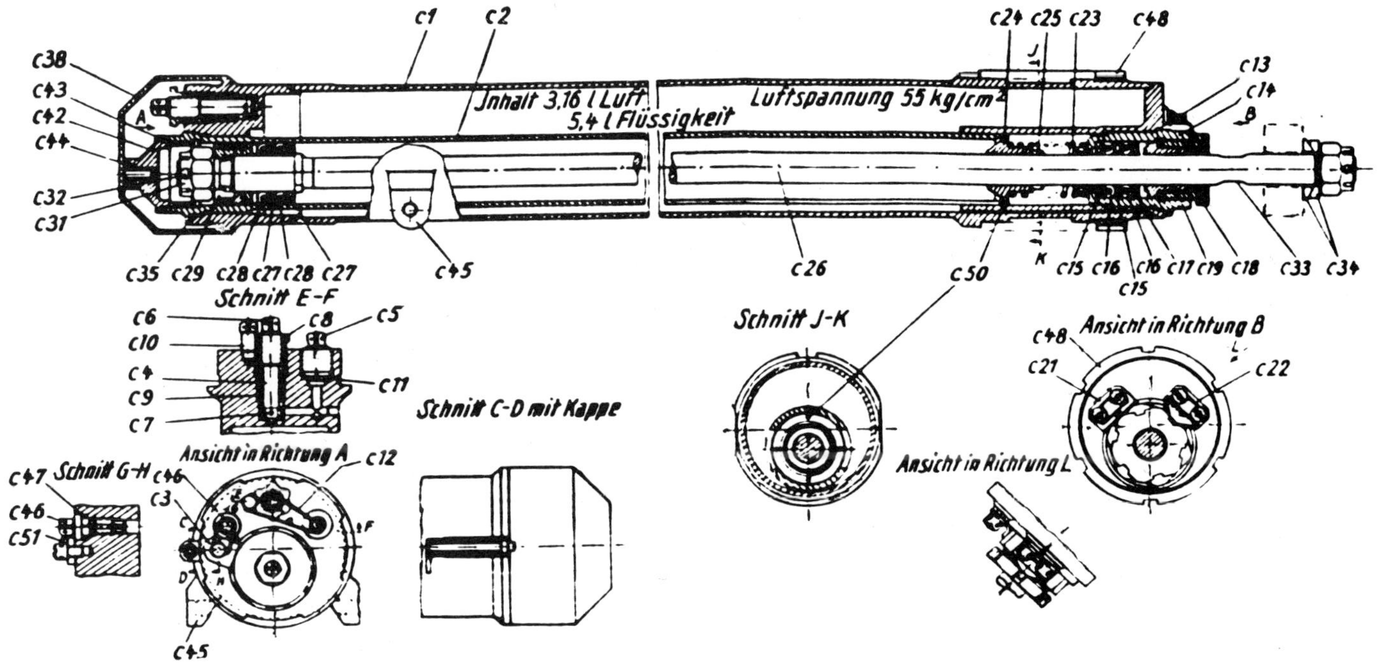

G

1	Air chamber	15	Spacer	32	Cotter pin	52	Crown nut
2	Compression cylinder	16	U-cup	33	Flattening	53, 54	Locking ring
3	Safety lock	17	Thrust collar	34	Washer		
4	Air valve	18	Stuffing box	35	Sealing ring		
5	Screw plug	19	White metal	38	Protective cap		
6	Spindle	21, 22	Safety locks	42	End nut		
7	Stop ring	23	Compression spring	43	Locking washer		
8	Gland-type fitting	24	Valve seat	44	Check valve		
9	Sealing ring	25	Flow apertures	45	Cams		
10	Locking screw	26	Piston rod	46	Screw		
11	Sealing ring	27	Thrust ring	47	Sealing ring		
12	Wire	28	U-cup	48	Dome nut		
13	Sealing ring	29	Rotating band	50	Borings		
14	Base bushing	30	Piston nut	51	Wire		

A new feature was the barrel recuperator located above the barrel, which used compressed air to return the barrel to firing position at the end of recoil. The device consisted of an air cylinder, a compression cylinder and the piston with piston rod. It was another example of the proven principle of liquid and air working together.

On June 1, 1927 the Heereswaffenamt (Army Ordnance Office) Testing Section issued Secret Command Matter No. 59/27, which called for the development of a new light field howitzer. The project was assigned Priority Level II, "most important work." Initial calculations and drawing of blueprints began in 1928, at first for a gun with a barrel length of L/25. Rheinmetall began production at the start of the 1930s.

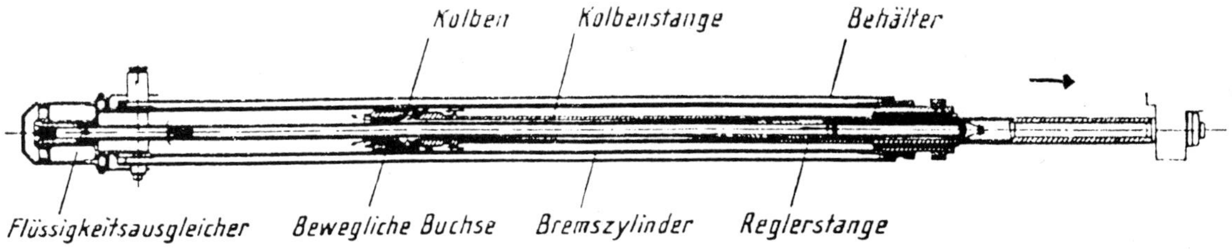

Bild der Rohrbremse im ersten Teil der Vorlaufbewegung

Bild der Rohrbremse während des Vorlaufes kurz vor Beendigung desselben

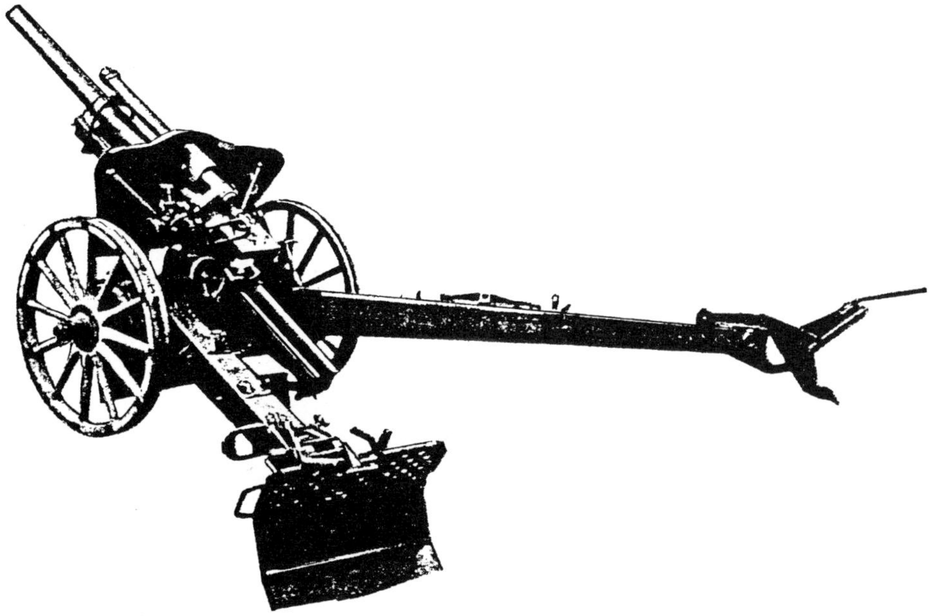

The tenth gun from the test and pre-production series of the leFh 18, still with the wooden spoked wheels of the leFH 16 and without a muzzle brake. However, the split-trail gun carriage, pneumatic recuperator and a thicker but lower gun shield were all significant advances. The decisive improvements had been made.

The upper carriage had a barrel brake in the cradle with water jacket and a fluid equalizer containing 6.2 liters of recoil oil. As well there was now also a barrel recuperator installed over the barrel. Filled with 3 liters of compressed air and 5.4 liters of liquid under 55 atmospheres of pressure, it returned the barrel to firing position immediately after firing. The cradle guide rail was extended to the rear, primarily on account of firing at greater barrel elevations of up to 42 degrees. When the barrel was under maximum stress the guide rails sometimes came flying out when the gun was fired. The pneumatic equilibrator within the gun shield (42 atmospheres) increased the forward weight of the barrel. A sixty-centimeter-long muzzle brake, which counteracted barrel recoil, was introduced in 1941. The stream of gases that poured out as kinetic energy at the muzzle after firing struck two chambers in the muzzle brake. The gases were redirected to the sides and rear, eliminating part of the recoil energy and thus reducing the

The 1936 production model leFH 18 (horse-drawn) featured new cast light-metal wheel discs with a diameter of 1.3 meters and removable resilient tires with a width of 10 centimeters. Between the reinforced spokes were twelve hand holes, and a lunette was attached to the hub for moving the gun by hand. (BA)

The tractor-drawn version of the leFH 18 waits at Lippstadt to be taken charge of by the 76th Artillery Regiment (Motorized) of the 1st Light Division. The wheel disc is the same size, but is fitted with an 11-centimeter-wide, solid-rubber tire, which was vulcanized on to the steel hub. The gun wears a recently-introduced camouflage finish. (BA)

recoil movement of the barrel. Muzzle velocity was simultaneously increased by 70 meters per second. The pusher-handle sliding-wedge type breechblock on the breech end of the barrel was fitted, not just with a locking mechanism, but a cartridge ejector as well. It was the beginning of automation. The sighting mechanism was simplified with the adoption of a follow-the-pointer sight mechanism, which consisted of a telescope mount reference pointer and a gun barrel reference pointer, so that the gun layer could set the elevation independent of the barrel and the 2nd Cannoneer had only to align the pointers. The emphasis of the innovations lay in the increased performance of the monobloc gun barrel with interchangeable charge liners, which soon resulted in an increase in firing range of 1,700 meters to 12.325 kilometers.

The result of all the improvements, which were fully developed and well-considered, was a weight increase of 400 to 500 kilograms, so that the field howitzer weighed almost two tons, strictly speaking already too heavy for transport by horse, especially later in Russia. Therefore the same model was built for transport by horse or motor vehicle, the only difference being 10 millimeter wider wheels with solid rubber tires for the tractor-drawn version. It was attached directly to a light or medium prime mover

I Battalion, 73rd Artillery Regiment (Motorized) of the 1st Panzer Division in Weimar received the new guns that same year and at once began conversion training and gun exercises. Important operational units were given priority in reequipping with the new weapons.

A photo taken from the brochure "Our Armed Forces." Participation by motorized artillery units equipped with the new leFH 18 (mot) in parades provided visible proof of the progressive building up of the Wehrmacht and its modernization. Seen here are two 5-ton prime movers (BN 9), Sonderkraftfahrzeug 6, one for the gun and even one for the platoon commander. (BA)

without a limber, a favorable arrangement that allowed a march speed of up to 40 kph on halfway-usable roads, equivalent to a day's march by a horse-drawn battery. The introduction of this field howitzer, on an order from the Commander-in-Chief of the Army on July 26, 1935, was announced on August 3 in the "General Army Communication." At the same time, however, the press was prevented from making any announcement. Rheinmetall had in fact come up with a good design which was adaptable and easy to maintain, extremely versatile, technically-balanced and durable. The barrel was good for 8,000 to 10,000 firings! As well there was the simplified process for aiming at moving targets, including tanks, a capability which the gunners, with their outstanding peacetime training, knew how to put to good use. No wonder that the leFH 18 with its total of 28 variants, many of which were mounted on captured and German-built chassis as self-propelled guns, became the standard gun of the Wehrmacht by 1945!

This leFH 18 has been moved into firing position in a meadow by its prime mover and unlimbered there during a peacetime exercise. The prime mover has remained on the scene, the gunners are opening the trails and preparing the gun for firing. Everything is still very casual. (BA)

A battery officer bawls out one of his gun crews during a maneuver. An ammunition-gunner rushes to the gun with the rammer, the breech is open. In contrast to the gunners, the gun layers always wore caps. On the right may be seen the limber pole for the horse-drawn limber. (BA)

Model 34 Sighting Mechanism

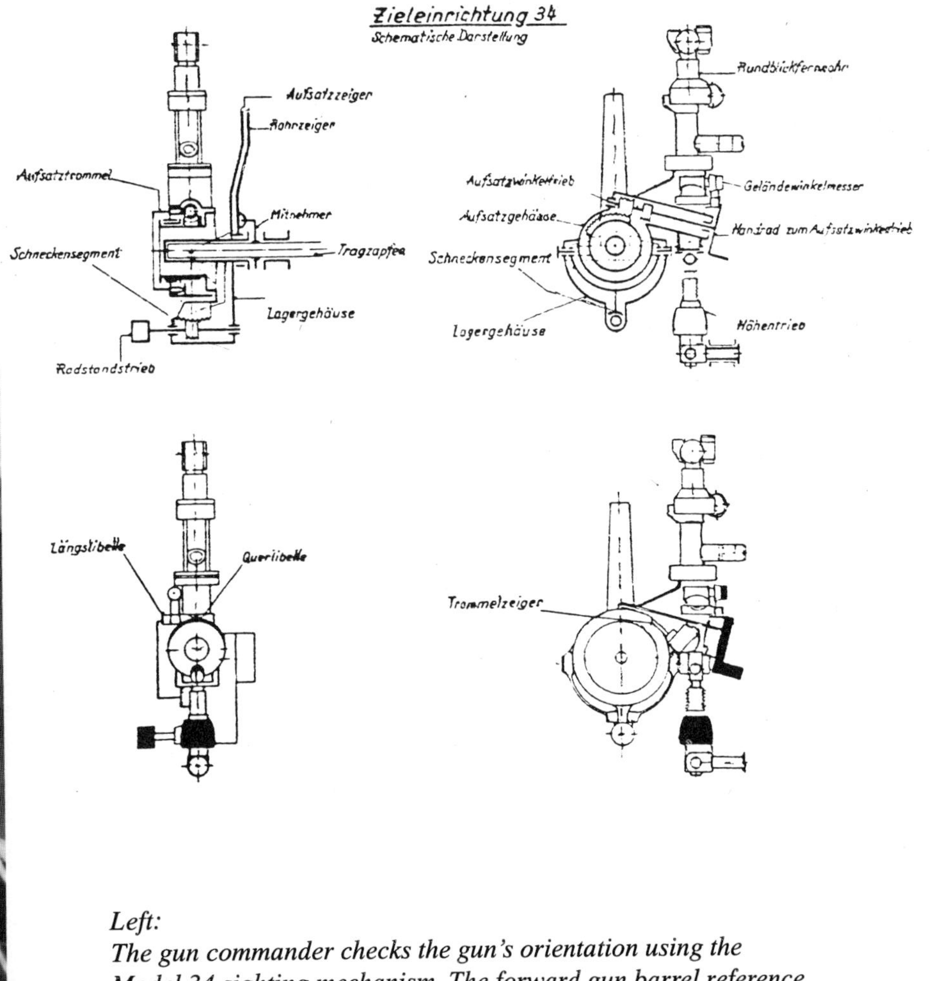

Left:
The gun commander checks the gun's orientation using the Model 34 sighting mechanism. The forward gun barrel reference pointer has yet to be adjusted, the barrel has not yet reached the desired elevation. The gun layer, who was also the deputy gun commander, was particularly well trained and was jointly responsible for the accuracy and effect of the fire. (BA)

Production and War Service

After 1939 training went on between campaigns, and not just for the replacement units. The conclusion and culmination of any training camp was live firing. During peacetime little emphasis was placed on winter firing, however this changed under the influence of war experience, as this photo of an leFH 18 (Pfzg) taken in January 1940 shows.

Units not employed in defending Germany's western border, those earmarked for an offensive role in the Western Campaign, which included the motorized artillery regiments, were rotated through the troop training grounds. Baumholder, Pfalz, May 3, 1940. (BA)

The German Armed Forces began the war with 4,862 leFH 18 light field howitzers. The total number of "leFH 18 on wheeled carriages" delivered between September 1939 and February 1945 totalled 6,933 guns, with the highest totals in 1940 and 1943. A total of 10,245 leFH 18/40 light field howitzers were delivered from March 1943 to March 1945, with 7,807 in 1944 alone. 662 of the leFH 18/2 version, which was mounted in the "Wespe" self-propelled gun, were built from February 1943 to May 1944, when production was phased out, while a total of 1,264 Sturmhaubitze 42s were constructed from March 1943 to March 1945. A total of 19,104 leFH 18 guns was therefore accepted, according to "Overview of the Equipment State of the Army." Rheinmetall and Krupp could not possibly produce these numbers as well as their other contracts; in 1942, therefore, they relocated all leFH 18 production to six firms in Pilsen, Altona, Elbing, Magdeburg, Dortmund and Borsigwalde. One must not forget that the mass of the artillery deployed by most of Germany's enemies, at least in the early years of the war, was inferior in caliber and often was not state of the art. This was one more reason for Hitler to insist on maintaining the silence concerning production figures until 1941. As well the leFH 18 was sold to Finland, Norway, Sweden, Spain, Hungary and Slovakia. The leFH 18 M as well as the 18/39 and the 39 were also purchased by the Netherlands.

The leFH 18/40 was an emergency solution brought about by the pressure of the war situation. In view of the large number of available carriages for the 75mm Pak 40 by Rheinmetall-Borsig and the great demand for the leFH 18, production technicians combined the howitzer's upper carriage with a slightly-modified anti-tank gun lower carriage for reasons of simplicity and standardization. The resulting hybrid was in principle the same weapon as the leFH 18 M or 18/39. The opportunity was taken to install several new features: doubled axle suspension through the use of torsion bars, which were automatically disengaged when the trails were spread; relocation of the elevating mechanism and the trigger from the right to the left side of the gun, to the gun layer's position; installation of a more effective muzzle brake; and a widening of the wheel hubs. The reduction of 240 kilograms in firing position also reduced the gun's stability, even though the lower muzzle height, which was favorable for engaging tanks, had the opposite effect. There was a slight improvement in the range or traverse to 60 degrees.

An leFH 18 under the command of a Leutnant employs direct fire to break resistance in the streets of Sedan, 1940.

According to D 97/1 Secret of April 12, 1945, there was a total of 24 different versions of six types of leFH 18.

5-1001 Light Field Howitzer 18 (M) (Horse-Drawn) – with muzzle brake
5-1002 Light Field Howitzer 18 (M) (Tractor-Drawn) – with muzzle brake
5-1016 Light Field Howitzer 18 (Horse-Drawn) – without muzzle brake
5-1020 Light Field Howitzer 18/40 (Horse-Drawn)
5-1032 Light Field Howitzer 18/39
5-1037 Light Field Howitzer 18/40 (Tractor-Drawn)
5-1040 105mm Assault Howitzer 42 – lFH 18 with jacket cradle and muzzle brake on Sturmgeschütz III
5-1041 105mm Assault Howitzer 42/1
5-1042 Light Field Howitzer 18/40/1 (Sf) – on Panzer IV
5-1043 105mm Assault Howitzer 42/2 – on Sturmhaubitzwagen 638/14
5-1044 Light Field Howitzer 18/40/3 (E) – on armored train
5-1045 Light Field Howitzer 18/1 (Sf) – on GW IV B
5-1046 Light Field Howitzer 18/2 (Sf) – on GW II "Wespe"
5-1048 Light Field Howitzer 18/3 (Sf) – on GW B 2 (f)
5-1049 Light Field Howitzer 18/4 (Sf) – on Lorraine-Schlepper (f)
5-1054 Light Field Howitzer 18/40/2 (Sf) – GW III/IV
5-1055 Light Field Howitzer 18/6 (Sf) – "Heuschrecke"
5-1061 Light Field Howitzer 18 (Tractor-Drawn) – for prime mover
5-1063 Light Field Howitzer 18/40/6 (Sf) – on GW 634/6
5-1065 Light Assault Howitzer 42/3 – on GW 638/13 (recoilless)
5-1066 Light Field Howitzer 18/40 in carriage 396 (r)
5-1067 Light Field Howitzer 18/40 in carriage 445 (r)
5-1087 Light Field Howitzer 18/40/4 (Sf) – on GW 638/19
5-1089 Light Field Howitzer 18/40/5 (Sf) – on GW 638/21

In Action – leFH 18 Horse-Drawn

The 5th Artillery Regiment encountered fleeing members of the civilian population for the first time on crossing the Seine near Mery on June 15, 1940.

Many units remained in France after the conclusion of the Western Campaign in 1940. Most of the artillery was positioned in makeshift positions on the Atlantic coast, until the construction of bunkers began in 1943.

Other artillery battalions rolled in countless trains through the Rhineland to their home military districts or continued on into the Generalgouvernement (German-occupied Poland), mostly horse-drawn units like the one seen here.

In the long run, however, the carriage trails proved too weak, especially when the gun was fired at high angles of elevation or when the sixth charge was used to take advantage of the gun's outstanding range. This problem disappeared when the weapon was mounted on a self-propelled carriage. Production began at the beginning of 1943 – between Stalingrad and Kursk; the first ten howitzers were delivered in March of that year and by the Battle of Kursk the figure had reached 418. In total the leFH 18/40 made up more than half of all light field howitzers produced.

Furthermore, the Rheinmetall, Krupp and Skoda Firms produced a number of advanced designs and prototypes before the war ended; their goals were increased firing range and a change from wheeled to

(Continued on page 44)

Stoff-gliederung 5 | **Geheime Kommandosache!** | Blatt G 116

leFH 18

Dringl.-St.: (Sf s. Kfz Bl. G 361–363 u. G 365–367)

Technische Daten:

Schußweite 10 675 m m. FH Gr F 12 325 m (nur m. Mündgsbr.)
Rohrweite 10,5 cm
Rohrlänge in Kalibern L/28 2,941 m
Geschoßgewicht (Sprenggranate) 14,81 kg

Größte V° 470 m/sek Sprgr.; 540 " Sprgr. F
Mündungswucht 167 mt
Zahl der Ladungen 6 + F bei MB
Gewicht in Feuerstellung 2 040 kg Besp.; 2065 kg Kfzg
Gewicht in Fahrstellung 3 490 " " ; 9800 " " einschl. Zgkw.
Richtfeld Seite 56° Höhe −5 +42°
Schuß/min 4–6
Geschätzte Lebensdauer eines Rohres 10–12 000 Schuß
Durchschl.-Lstg.: auf 100 m = 56 mm, auf 1500 m = 44,5 mm (5. Ldg. 10 cm Pz Gr)
" " = 64 " ; " " = 49 " (6. Ldg. " ")
" " = 70 " ; " " = 70 " L 10 cm Gr 39 rot)
" " = 100 " ; " " = 100 " (10 cm Gr 39 rot HL/C)

Geschütz-Verwertung (Mündungswucht mt / Geschütz-Gewicht t) ...... 85,5 mt/t [bei Geschützen i. gepanz. Fahrzeugen ohne Bedeutung]

Rohstoffbedarf	Fe	Mo	Cr	W	Mg	Sn	Cu
f. 1 Stck. i. kg	6190	2,8	3,567		0,8	0,756	7,068
	Al	**Pb**	**Zn**	**Ni**	**Kautschuk**		
f. 1 Stck. i. kg	0,187	0,240	0,956		56,5 + 0,3		

Preis RM 16 400.–	Durchschn. Fertigungszeit 6 Monate	Arbeitsstunden 3200

Fertigungsfirmen:
wird nicht mehr gefertigt

Geschoß:

F.H.Gr.
F.H.Gr. Stg.
F.H.Gr. 38
F.H.Gr. 38 Stg.
F.H.Gr. 40 Nb
F.H.Gr. Nb
F.H.Gr. 38 Nb
F.H.Gr. Br.
F.H.Gr. (Üb)
F.H.Gr. Stg. (Üb.T.)
F.H.Gr. (Üb.T.)
F.H.Gr. 38 (Üb.T.)
F.H.Gr. 38 Stg. (Üb.T.)

10,5 cm Sprgr. 43
*) F.H.Gr. F
10 cm Pzgr.
10 cm Pzgr. (Üb.)
10 cm Pzgr. 39 TS u Al
10 cm Sprgr. 42 TS
10 cm Gr. 39 rot HL/B
10 cm Gr. 39 rot HL/C
10 cm Weiß-Rot-Geschoß
F.H.Gr. 40 Deut
F.H.Gr 39 als Sprgr
F.H.Gr 38 Kh als Sprgr.
F.H.Gr. 35
10,5 cm Lt Gs
10,5 cm Br Gs

*) darf nur aus Geschützen m. Mündgsbr. verschossen

Specification sheet issued by the Reich Minister for Armaments and War Production on November 1, 1943. As well as data concerning the technical aspects of the leFH 18, the document provided information concerning materials usage and costs. 28 different types of shell could be fired from the leFH 18, an indication of its versatility.

Left:
III Battalion, 96th Artillery Regiment, part of the 44th Infantry Division, in parade formation in early 1941, prior to the start of the campaign against Russia. It was farewell to an almost peaceful interlude.

Right:
Early April 1941. The Number 4 gun of a horse-drawn battery crosses the Shipka Pass between Gabrovo and Kasanlik in Bulgaria, the most important pass in the Balkan Mountains, elevation 1,300 meters. In the background is an Orthodox monastery, which was much fought over in 1877-78. (BA)

Left:
Summer 1941. Number One gun of an artillery battery immediately after firing. Judging by the stockpile of ammunition, the battery is taking part in a major fire mission. The loader waves for the next round, for which the cartridge is just being prepared. (BA)

A gun emplacement shrouded in morning fog, Southern Russia, early autumn 1942. The gun commander has raised his hand, the signal to fire the gun, and the crew has stepped back, while the 3rd gunner is already picking up the next shell. A major battle meant grueling work for the gun crew. (BA)

Right:
The divisional artillery of the panzer units played a major part in the advance on Kharkov, which was recaptured on March 14, 1943. The rapid and successful counterattack had an invigorating effect on the troops.

Left:
The German retreat in December 1941, with the troops still wearing summer uniforms in the bitter cold, exhausted men and horses and resulted in considerable losses. Heavy, horse-drawn vehicles were especially hard hit; many became stuck and had to be abandoned. Even saddle horses had to be led.

Right:
The first proper winter combat clothing reached artillery units in Northern Russia in the winter of 1943. Noticeably-larger draught horses are now being used; the limber is loaded with feed for the horses, forcing the gunners to walk.

Table of organization for a horse-drawn Light Field Howitzer 18 battery belonging to an infantry division, November 1, 1941.

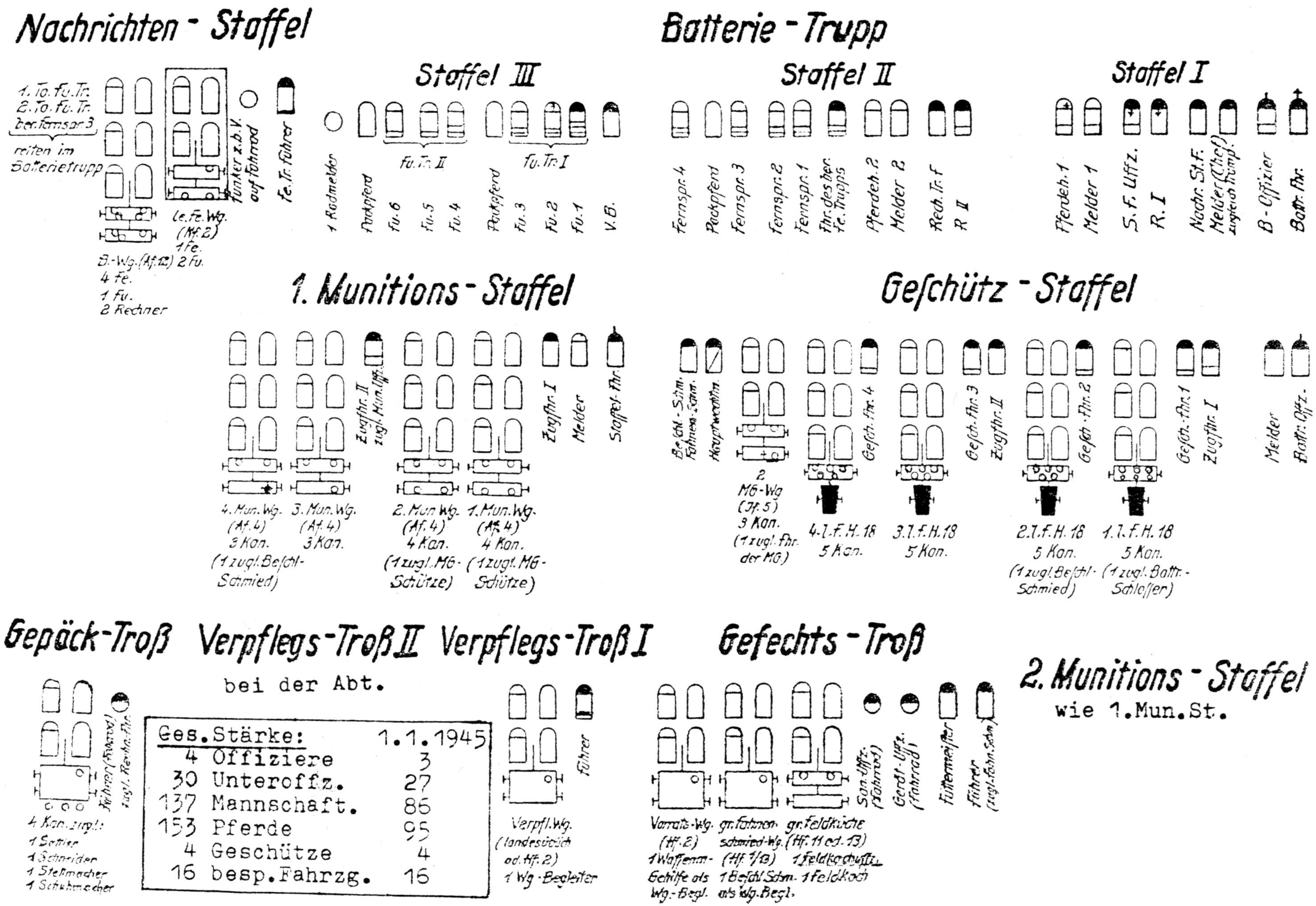

Bosnia, Yugoslavia, late October 1943. Using both hands, the 2nd gunner pulls on the lanyard to release the firing pin, which fired the gun. For some time the leFH 18 (Tractor-Drawn) was also built with wooden wheels, with vulcanized solid-rubber tires instead of metal ones. (BA)

Stoffgliederung 5	Geheime Kommandosache! 10,5 cm Geb H 40	Blatt G 112

Dringl.-St.: SS

Technische Daten:

Schußweite 12 625 m

Rohrweite 10,5 cm

Rohrlänge in Kalibern L/30 = 3,150 m

Geschoßgewicht 14,81 kg

Größte V_0 570 m/sek

Mündungswucht 245 mt

Zahl der Ladungen 7

Gewicht in Feuerstellung 1656 kg

(Gewicht in Fahrstellung) 4 Fahrlasten zu je ~650 Kg

Richtfeld Seite 50° Höhe -5 +70°

Schuß/min 4-6

Geschätzte Lebensdauer eines Rohres 8-10 000 Schuß

Durchschl.-Lstg.: 100 mm starke Panzerplatte m. 10 cm Gr. 39 rot Hl/C (s. Vbm I)

Geschütz-Verwertung (Mündungswucht mt / Geschütz-Gewicht t) 148 mt/t [bei Geschützen i. gepanz. Fahrzeugen ohne Bedeutung]

Rohstoffbedarf	Fe	Mo	Cr	V	Mn	Sn	Cu
f. 1 Stck. i. kg o. Räder u. Zieleinrichtg	6520	0,8	39,4	1,5	21,5	0,16	5,43
	Al	**Pb**	**Zn**	**Ni**	**Kautschuk**		**Si**
f. 1 Stck. i. kg		0,11	2,23	4,0	7,1		6,0

Preis RM 54 850.-	Durchschn. Fertigungszeit 8 Monate	Arbeitsstunden 8500

Fertigungsfirmen:

Gebr. Böhler, Kapfenberg

Geschoß:

F.H. Gr. Al 14,81 kg
F.H. Gr. Buntr. 14,81 "
10,5 cm Spr. Gr. 43 14,81 kg
10,5 cm Lt. Ga
F.H. Gr. 38 Al
10 cm Gr. 39 rot Al/A, B, C 12,3 12,1 12,35 kg

This outstanding design, more mobile than the leFH 18 and costing the same as the 100mm sKanone, was supposed to replace the light field howitzer in the mountain artillery from 1944.

Right:
On July 15, 1941 the 1st and 4th Mountain Infantry Divisions set off near Bar to attack the Stalin Line. Three weeks later they had reached the Dniepr River. The 3rd battalions of the 79th and 94th Mountain Artillery Regiments supported the advance by bombarding known enemy positions.

Left:
While moving to more distant firing positions in February 1944 the gunners and even the drivers walked to spare the teams, especially since heavy horses tired more easily. The gun wears a whitewash winter finish; all the soldiers are wearing winter clothing. (BA)

Right:
A battery passes through a burning Russian village during Army Group Center's retreat to the "Panther Position" in January 1944. The gunners have slung their carbines over the barrel of the gun in a scene typical of this period. (BA)

A winter field emplacement belonging to the 7th Battery of the 178th Artillery Regiment, part of the 78th Assault Division, in the "Panther Position" west of Novoye Selo (north of Orsha). Taken in December 1943 following the vicious fighting of the third highway battle, the photograph gives the impression of chilly isolation.

Right:
A ten-horse team, and all white horses at that. Quite an unusual arrangement until one considers the load: the cross-country hauling of an leFH 18 M with wooden wheels and solid rubber tires on an leFH 16 limber, and a large number of gunners riding on the vehicles. It was pure improvisation on account of the desperate situation. (BA)

Left:
Although the artillery had been playing an increasingly prominent role in the German defense since 1943, by 1944-45 ammunition had long since become scarce on account of raw material shortages and production difficulties. Permission to fire was only given on express order and in the most dire emergency. Daily ammunition reports were used to keep track of ammunition stocks and usage.

The entire gun crew had to pitch in to clean the battery's guns of powder residue, which could affect accuracy, here during a pause in the defensive fighting against the Americans in the west in 1944.

The guns of an artillery regiment of a division moved from France to the Eastern Front. Although it is equipped with the leFH 18 M, the guns have the wooden wheels of the leFH 16 – not an ideal improvisation, but a practical one. The gun carriage, whose lower shield has been folded down, even though it is not necessary, was made lighter as a result of this modification. (BA)

During a pause in the fighting gunners dry their socks over a small fire after a cold, damp night on the Oder Front near Frankfurt. The fire is kept low so as not to betray their position. This is a very realistic picture of the daily life of a gunner. The leFH 18 has the standard replacement wooden wheels with vulcanized solid rubber tires intended for motor transport. Three pieces of sheet metal were riveted to the wheel to keep the rubber tire from slipping off.

The Tractor-Drawn leFH 18

Driving courses and demonstration drives with the new 3-ton prime mover were carried out in the assembly areas in June 1941 shortly before the beginning of the invasion of Russia. Developed by the Reichswehr, this half-tracked vehicle proved a great success in all its variants. (BA)

The German advance went on inexorably in the area north of Smolensk on July 11, 1941. The objective of the advance by the panzer and motorized infantry divisions was to close the pocket around the city. The 3-ton prime mover (Sd.Kfz. 11) HKL 6 stood the test as a standard mode of transport for the leFH 18; the 5-ton prime mover was not always required. (BA)

Table of organization of an leFH 18 battery (motorized).

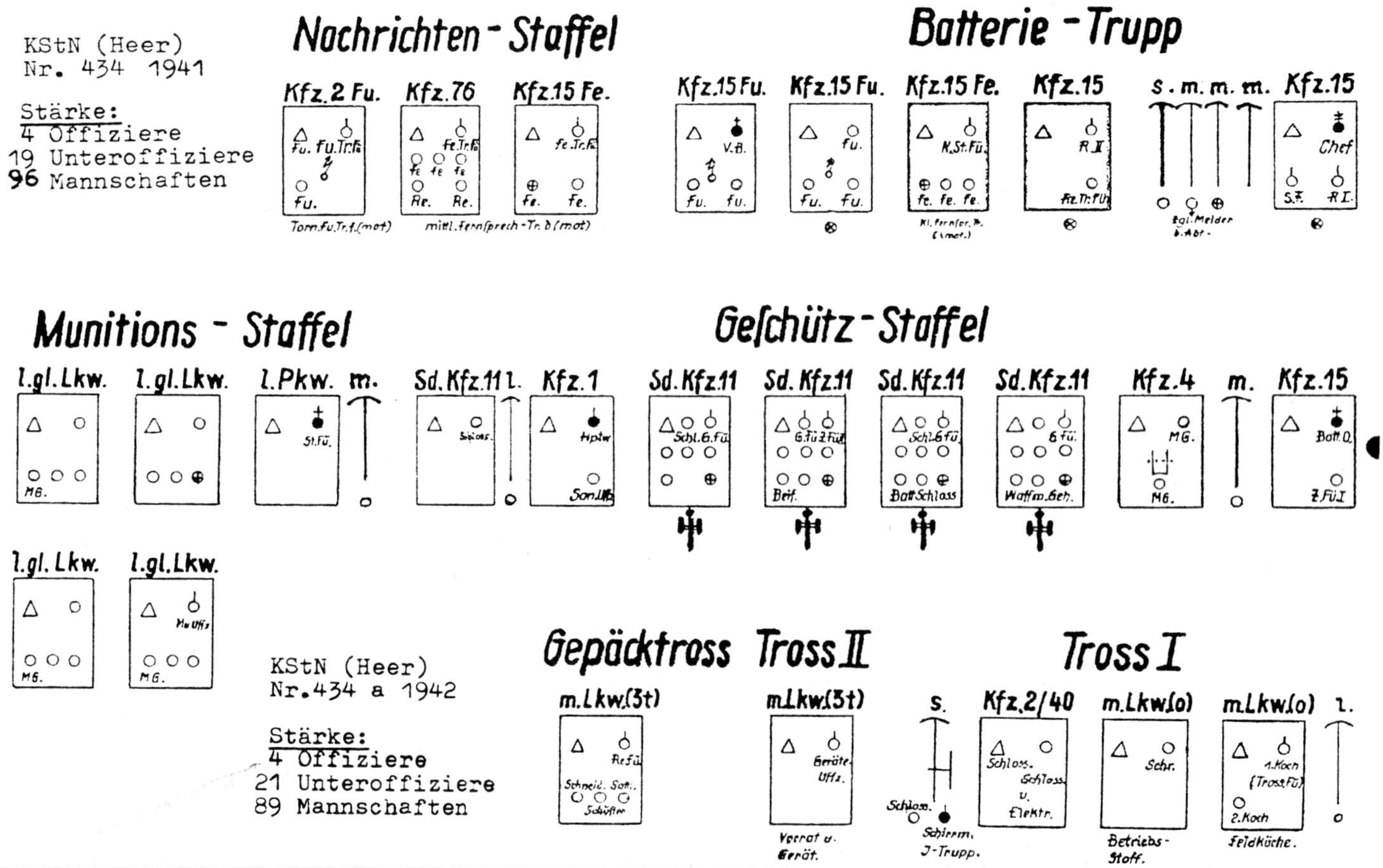

Motorized artillery batteries were more rational and considerably more effective than their horse-drawn counterparts. Their radius of action was six to ten times greater, with a considerably lower expenditure in personnel and vehicles, to say nothing of greater speed and mobility in action. Maintaining a large number of horses was a great burden on the mass of artillery batteries. A motorized battery required eight less NCOs and 41 enlisted men; nevertheless it had no personnel reserve. In addition the battalion provided three munitions trucks for each battery. (BA)

In the German Africa Corps the motorized field artillery was an important weapon alongside the armored forces. Here an leFH 18 unlimbers in completely open terrain, while two gunners loosen bolts and catches for the cradle lock before spreading the trails. (BA)

Right:
The leFH 18 (mot) is limbered to the 3-ton prime mover again. The 2nd gunner hangs the often-necessary pickaxe on the upper shield. There were no concealed emplacements in the desert.

Left:
The gun layer raises his hand, indicating to the gun commander that the weapon is ready to fire, and steps back. Sighting through the open sight port, he has taken direct aim at the distant, specified target. The gun commander, standing beside the gun layer, waits for the dust to settle before giving the order to fire.

Right:
A battery of the 44th Artillery Regiment (Motorized) of the 15th Panzer Division fires at advancing enemy tanks from an open field emplacement. The regiment destroyed more than 100 British tanks at Sidi Rezegh on November 23, 1941. The platoon and gun commanders observe the fall of the shot. Four gunners struggle to hold the folding trail spade in the stony ground. (BA)

Early October 1941. A breakthrough by the enemy forces a battery to open fire unexpectedly at short range. Prepared shells lying on a tarpaulin and the many empty ammunition boxes are an indication of the tension and the fierceness of the battle. (BA)

An leFH 18 carries out the first fire mission in November 1941 in Northern Russia. The 2nd gunner pulls the lanyard; the gun layer, wearing the cap, – his role did not call for a steel helmet, –has stepped back. The gun commander holds his firing tables in his left hand. (BA)

Left:
A battery sets up in new positions near Rzhev, January 1942. The gun commander checks the work of the gun layer while the gunners make the weapon ready to fire. On account of the frozen ground the trails rest on their ice spurs. (BA)

Right:
During a pause in the firing gunners eat hot food delivered by the Senior NCO. The gun has received a coat of whitewash, but the men are still without winter clothing, a realistic picture. (BA)

Below:
A gun emplacement comes under heavy fire during Russian attacks on the "Panther Position" in 1943. As a rule, during important fire missions the guns had to continue firing no matter what the circumstances. The 2nd gunner is about to fire the gun.

Left:
The first artillery battalions equipped with the RSO Tractor (top speed 17 kph) and the leFH 18/40 with larger muzzle brake reached the field during the buildup prior to the Battle of Kursk. As the RSO had seats for only two, leFH 18 limbers were hitched between the tractor and gun for the crews to ride on. The tractor was capable of pulling 2 tons. These batteries often had six guns. (BA)

Right:
An leFH 18/40 in a prepared position in Southern Russia, autumn 1943. The gun's scanty camouflage of branches has been removed and the gun layer is taking aim. The weapon's similarity to the 75mm, 67.2mm and 88mm anti-tank guns and the Russian 76.2mm Model 42 cannon is obvious. Firing at high elevations was hard on the trails. (BA)

Left:
The 3rd gunner pushes a shell into the open breech with the rammer. An ammunition-gunner is ready with the cartridge. The gun carriage has wooden wheels with rubber tires, an interim solution. The sighting mechanism is in the extended position for direct aiming, an indication that an attack by enemy tanks is expected. Army Group Center, Russia, summer 1944. (BA)

On March 5, 1942 Hitler requested a combination of the leFH 18 with the carriage of the 75mm Pak 40. Rheinmetall-Borsig of Düsseldorf was charged with its development the same day. The weapon was ready for production by October 15, 1942. The first examples were delivered to the Wehrmacht in March 1943. Quantity production began in April 1943, followed by deliveries to the units on September 15 of the same year.

Stoffgliederung	Geheime Kommandosache!	Blatt
5	le F H 18/40	6116 II

Dringl.-St.: SS

Technische Daten:

Schußweite m. F H Gr = 10675 m, m. F. H. Gr. F. = 12325 m

Rohrweite 10,5 cm

Rohrlänge in Kalibern L/28 = 2,941 m

Geschoßgewicht 1) F. H. Gr 14,81 kg
2) F. H. Gr. F. 14,81 "

Größte V_0 1) 470 m/sek bei FH Gr
2) 540 m/sek " " " F

Mündungswucht 167/220 mt

Zahl der Ladungen 5+1+1

Gewicht in Feuerstellung 1900 kg

Gewicht in Fahrstellung 2900 kg

Richtfeld Seite 60° Höhe −5/+42°

Schuß/min 6–8

Geschätzte Lebensdauer eines Rohres ~ 10 000 Schuß

Durchschl.-Lstg.: m 10 cm Gr. 39 rot Hl/C = 100 mm

Geschütz-Verwertung (Mündungswucht mt / Geschütz-Gewicht t) 116 mt/t (bei Geschützen i. gepanz. Fahrzeugen ohne Bedeutung)

Rohstoffbedarf	Fe	Mo	Cr	W	Mg	Sn	Cu
f. 1 Stck. i. kg	3991	22,9	6,9	1,6		0,24	5,1
	Al	Pb	Zn	Ni	Kautschuk		
f. 1 Stck. i. kg		0,2	1,1				

Preis RM	Durchschn. Fertigungszeit Monate	Arbeitsstunden

Fertigungsfirmen:
Schichau, Elbing
Menck & Hambrock, Hamburg
Krupp, Markstaedt

Geschoß: s. Bl. 6116

The 2nd gunner closes the breech in preparation for firing. The gun layer is already covering his ears, while the ammunition-gunner readies the next cartridge. A low silhouette and wide range of traverse distinguished the leFH 18/40. (BA)

December 8, 1944. An leFH 18/40 battery fires at enemy assembly areas somewhere in Western Europe. The gun is maintaining a high rate of fire.

Stoffgliederung 5	Geheime Kommandosache! le FH 18/40/2 (SF)	Blatt 6 116 IIa

Dringl.-St.:

Technische Daten:

Schußweite 10675 m

Rohrweite 105 mm

Rohrlänge in Kalibern L/28 2941 mm

Geschoßgewicht 14,81 kg

Größte V_0 470 m/sek

Zahl der Ladungen 6 + Fernladung

Gewicht in Feuerstellung 1150 kg

Gewicht in Fahrstellung (SF) 25000 kg

Richtfeld Seite (m.Pilz) 360 Höhe -5 + +42°

Schuß/min 6

Geschätzte Lebensdauer eines Rohres 8-10000 Schuß

Durchschl.-Lstg.: 100 mm. mit HL-Gr. auf 1000 m / 60°

Geschütz-Verwertung (Mündungswucht mt / Geschütz-Gewicht t) .... [bei Geschützen i. gepanz. Fahrzeugen ohne Bedeutung]

Rohstoffbedarf f. Stck. i. kg	Fe	Mo	Cr	W	Mg	Sn	Cu
f. Stck. i. kg	Al	Pb	Zn	Ni	Kautschuk		

Preis RM -	Durchschn. Fertigungszeit Monate	Arbeitsstunden

Fertigungsfirmen: Wird nicht mehr gefertigt	Geschoß:

Following unsuccessful attempts to install the leFH 18/40 on the 3-ton prime mover in the summer of 1943, the weapon was mounted on captured French Hotchkis and F.C.M. tanks as an armored howitzer. The leFH 18/40 was also mounted on the chassis of the Panzer II ("Wespe") and Panzer IV ("Heuschrecke"), which were then called "Geschützwagen" (self-propelled guns), abbreviated GW. The above specification sheet is for the self-propelled version of the leFH 18/40.

Stoff-gliederung	Geheime Kommandosache!	Blatt
5	le F.H. 18/2 f. sf	G 116 III

Dringl.-St.:

Technische Daten:

Schußweite 10650 m
Rohrweite 10,5 cm
Rohrlänge in Kalibern L/28 = m
Geschoßgewicht (Sprenggranate) 14,81 kg

Größte V_0 470 m/sek m. Spr.Gr.
Mündungswucht 167 mt
Zahl der Ladungen 6 u. Fernladung
Gewicht in Feuerstellung } 11500 kg
Gewicht in Fahrstellung }
Richtfeld Seite ±20° Höhe -5+ +42
Schuß/min 6
Geschätzte Lebensdauer eines Rohres 12000 Schuß
Durchschl.-Lstg.: 100 mm auf 1000 m bei 60°

Geschütz-Verwertung (Mündungswucht mt / Geschütz-Gewicht t) 14,4 mt/t (bei Geschützen i. gepanz. Fahrzeugen ohne Bedeutung)

Rohstoffbedarf	Fe	Mo	Cr	W	Mg	Sn	Cu
f. Stck. i. kg							
	Al	Pb	Zn	Ni	Kautschuk		
f. Stck. i. kg							

Preis RM 11 300.–	Durchschn. Fertigungszeit 4 Monate	Arbeitsstunden 2200

all-round carriages, definitely the right direction for future development but overtaken by events. As well the military demands on the field howitzer changed with experience. For example, they were supposed to have an anti-tank capability as well as be able to fire at extreme angles of elevation on account of the many forest battles, a role to which heavy mortars were much better suited. The Field Howitzer 18/42 by Krupp, with a barrel length of 31 calibers, was actually just an improved model of the leFH 18 M L/28 with a lengthened barrel and improved range (12,700 meters), but like the 18 M it weighed 2,035 kilograms. The leFH 42 was an improved version of the leFH 18/40 and the first light howitzer to exceed a firing range of 13 kilometers. Weight was reduced by 170 kilograms to 1.63 tons. Innovations included wider wheel hubs, electrical firing and barrel recuperator beneath the barrel in the cradle, which allowed the upper carriage to be kept quite low in profile.

On the other hand the designs by Krupp and Skoda for the leFH 43 incorporated an outrigger-type lower gun mount based on the example of the 88mm Flak and 88mm Pak 43 for all-round firing. This was a tremendous innovation, one which was picked up by France and Sweden after the war. (Such a carriage was suitable for transport by gun tractor only.) Krupp's Model I, with a barrel length of 28 calibers, achieved a muzzle velocity of 720 m/sec and a maximum range of 15,000 meters, exceeding anything extant. The Krupp Model II had a barrel length of 35 calibers, which approached that of the 100mm cannon. Firing a rocket-boosted projectile, it was capable of firing ranges as great as 16,500 meters.

The barrel of the 105mm lFH 18 as used by the "Wespe" self-propelled gun.

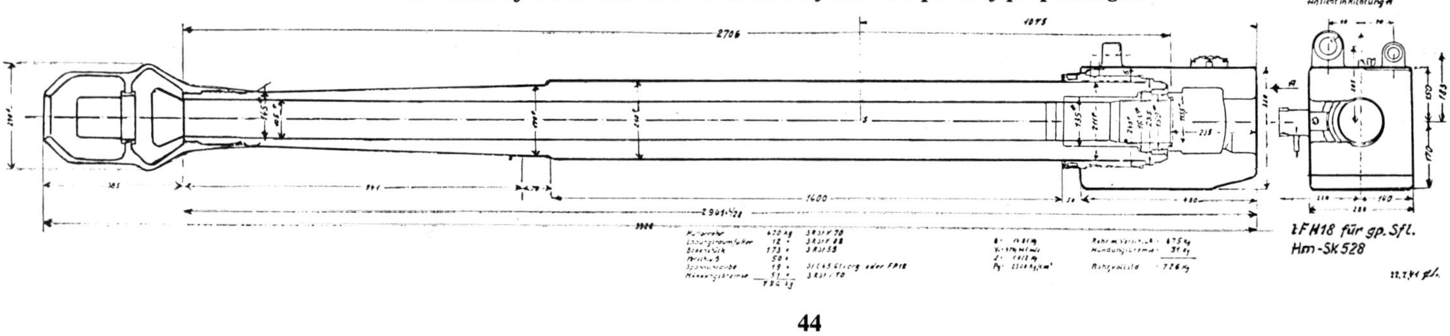

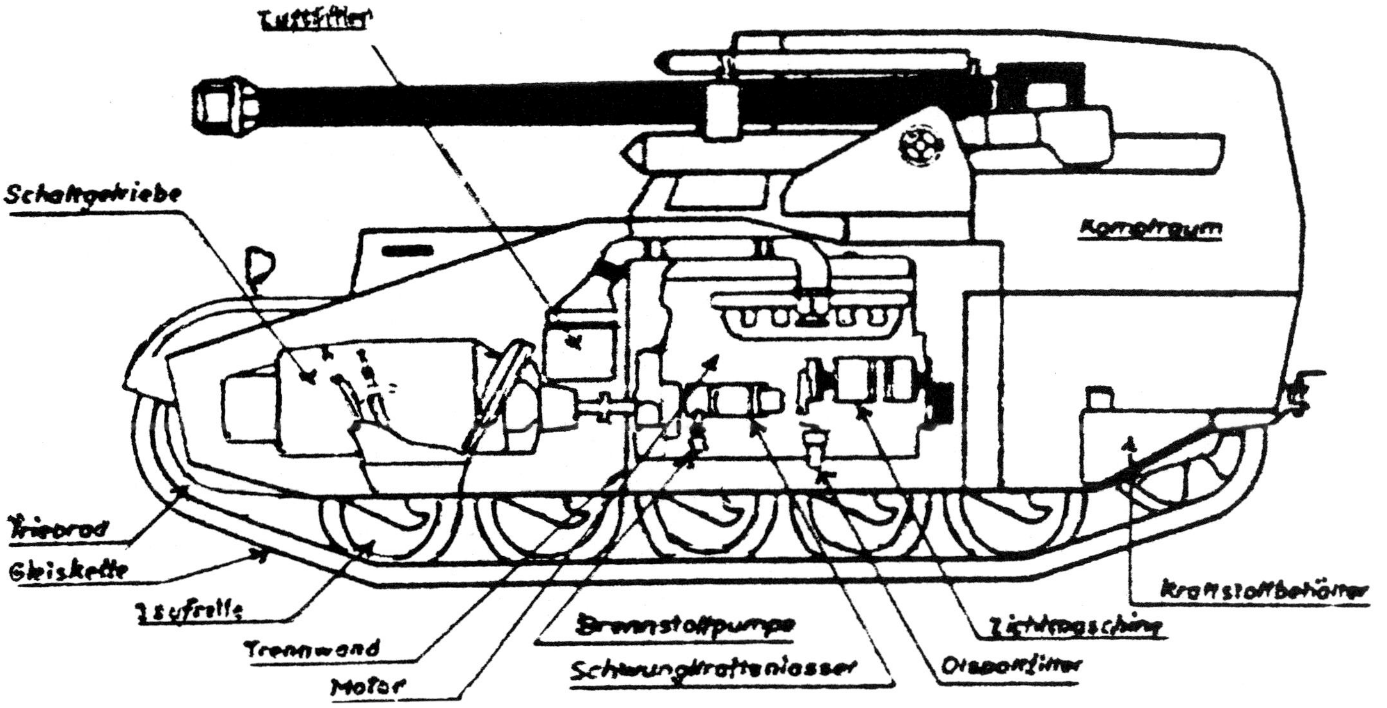

The superstructure of the Geschützwagen II for the leFH 18/2 (Sf) reveals a modern design, equivalent to present day standard.

This "Wespe" has just driven into firing position, the driver is still in his place. The gun commander checks the terrain in front of the vehicle, while the gunners prepare the gun for firing. (BA)

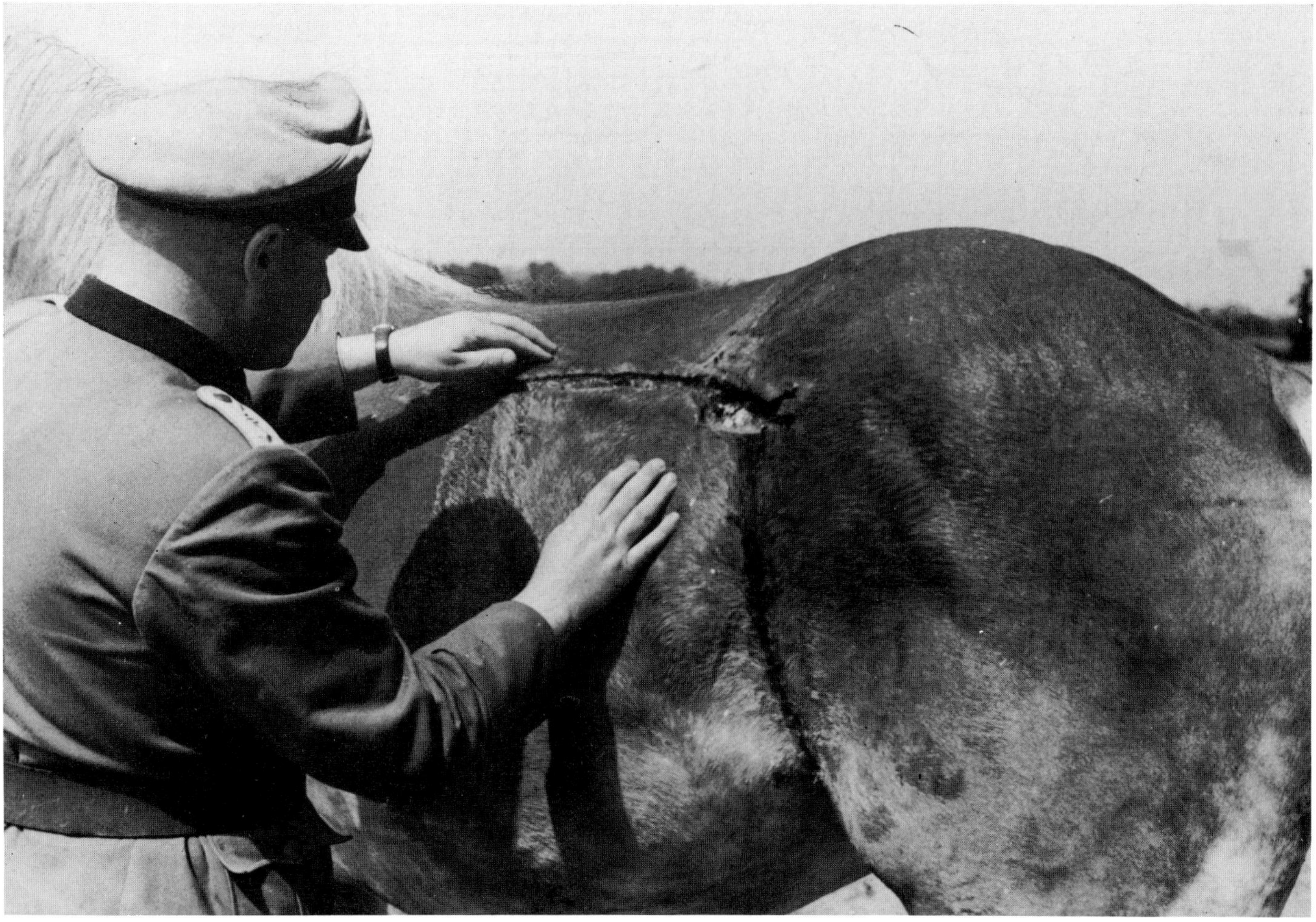

Most light artillery would have been incapable of movement without able veterinary officers.

EPILOGUE

It turned out that it was possible to wring even greater performance from the basic light field howitzer design. In its early years, the postwar Bundeswehr still had a light field howitzer with a weight in firing position of 2.5 tons, a barrel length of 32 calibers, muzzle velocity of 640 m/sec and a range of 14,170 meters. It was an improved version of the American M-2A1 105mm field howitzer, now for motor transport only, however. All other considerations aside, guns weighing more than two tons were two heavy to be transported by horses, especially off-road. Since a tank cannon with a caliber of 105 millimeters had been adopted as a standard weapon (120mm today) the light field howitzer had become too small in caliber to be a standard weapon of divisional artillery and had been overtaken by the general improvement in weapon effectiveness. Furthermore it was too costly to produce, and the enormous increase in armored vehicles, including on the part of the enemy, limited its effectiveness considerably. Even though light field howitzers contested the First World War to a large degree and the Second World War almost exclusively, the difference between light and heavy howitzers has meanwhile been overtaken and the age of light field howitzers has become part of history!

Technical Data

Modell	10.5 cm le. FH 16	10,5 cm le. FH 18	10,5 cm le. FH 18 M	10,5 cm le. FH 18/40	10,5 cm le. FH 18/42	10.5 cm le. FH 42	10,5 cm le. F.H. 43 (Krupp I) Holm-lafette	10,5 cm le. F.H. 43 (Krupp II) Kreuz-lafette	10,5 cm le. FH 43 (Skoda)
Kaliber mm	105	105	105	105	105	105	105	105	105
Rohrlänge mm	2310	2941	3308 (m. M.)	2710	3255	2940	2941	3675	3456
Seelenlänge mm	.	.	.	.	.	.	.	.	.
Gezogener Teil mm	1878	2392	2392	2392	.	.	.	.	.
Verbrennungsraum mm	106,5	212	212	.	.	.	.	.	.
Züge Zahl	32	32	32	32	.	.	.	.	.
Züge Tiefe mm	.	.	.	1	.	.	.	.	.
Züge Breite mm	.	.	.	5.5	.	.	.	.	.
Felderbreite mm	.	.	.	.	.	.	.	.	.
Drall, Art und Winkel	zunehmender Rechtsdrall 4°—10°	zunehmender Rechtsdrall 6°—12°	zunehmender Rechtsdrall 5°—8°	zunehmender Rechtsdrall 5°—8°	zunehmender Rechtsdrall 6°—9°	.	.	gleichbleibender Drall, 9°	.
Abfeuerung	Schlagbolzen	Schlagbolzen	Schlagbolzen	Schlagbolzen	Schlagbolzen	elektrisch	Schlagbolzen	elektrisch	.
Rücklauf maximal mm	1270	1170	1150	1150	.	.	.	.	.
Rücklauf minimal mm	1230	1150	1150	1150	.	.			1550/1150/850 (3 Stellungen)
Richtmittel	Rundbl.-Fernrohr	Rundbl.-Fernrohr	Rundbl.-Fernrohr	Rundbl.-Fernrohr	Rundbl.-Fernrohr	Rundbl.-Fernrohr	Rundblickfernrohr	Rundblickfernrohr	Rundbl.-Fernrohr
Schwenkbereich°	4	56	56	56	60	70	360	360	360
Erhöhung°	−10 +40	−5 +42	−5 +42	−5 +42	−5 +45	−5 +45	−4 +70	−10 +70	−5 +75
Feuergeschwindigkeit S/min	6—8	6—8	6—8	6—8	6—8	6—8	6 8	6	6 8
Fahrstellung Länge mm	5000	6100	.	6150	.	.	.	.	.
Fahrstellung Breite mm	1650	1977	.	2110	.	.	.	.	.
Fahrstellung Höhe mm	1650	1880	.	1830	.	.	.	.	.
Gewicht Fahrstellung kg	2870	3490	3540	2900	.	.	ca. 2300	.	2200
Gewicht Feuerstellung kg	1525	1985	2040	1800	2035	1630			
V_0 maximal m/s	395	470	540	540	585	595	720	655	690
Kammerdruck atü	2150	2300	2300	2300	2500	.	.	2500	2600
Gebrauchsgasdruck m/t	118	167	220	.	.	.	.	.	.
Schußweite maximal m	9225	10675	12325	12325	12700	13000	15000	16500	15000
Zahl der Ladungen	5	6	6+1	6+1	7	7	7	7	7
Granaten	FH Gr. 10,5 cm Gr. 39 Rot Hl. Hl/A. Hl/B u. Hl/C	FH Gr. FH Gr. 38 Stg. 10 cm Pz.Gr. 10 cm Pz.Gr. Rot 10 cm Gr. 39 Rot FH Gr. Nb. FH Gr. 38 Nb.	wie 10,5 cm le. FH 18 dazu FH Gr. Fern	FH Gr. 38, 39, 40 u. 41 FH Gr. Br. FH Gr. Sp. Br. 10.5 cm Lt.Gs., Gr. 39 Rot Hl. Weiß-Rot Geschoß, Sprgr. 39TS u. 42TS. Minen Gr.	wie 10,5 cm le. FH 18/40	wie 10,5 cm le. FH 18/40	wie 10,5 cm le. F.H. 18/40	wie 10,5 cm le. F.H. 18/40	wie 10,5 cm le. FH 18/40

It was a long road from the leFH 16 to the leFH 18/2, spanning thirty years and two wars!

The Spielberger German Armor & Military Vehicles Series

Size: 8 1/2" x 11" 288 pages hard cover
over 460 photographs
ISBN: 0-88740-397-2 $39.95

Size: 8 1/2" x 11" 256 pages hard cover
over 240 photographs
ISBN: 0-88740-398-0 $39.95

Size: 8 1/2" x 11" 168 pages hard cover
over 200 photographs
ISBN: 0-88740-448-0 $29.95

Size: 8 1/2" x 11" 168 pages hard cover
over 200 photographs
ISBN: 0-88740-515-0 $29.95